Chemical Ventures Conference 2019

AA009311

Wilmington, Delaware, USA
23-24 April 2019

ISBN: 978-1-7138-2863-1

Printed from e-media with permission by:

Curran Associates, Inc.
57 Morehouse Lane
Red Hook, NY 12571

Some format issues inherent in the e-media version may also appear in this print version.

AIChE
120 Wall Street, FL 23
New York, NY 10005-4020

Phone: (800) 242-4363
Fax: (203) 775-5177

www.aiche.org

Additional copies of this publication are available from:

Curran Associates, Inc.
57 Morehouse Lane
Red Hook, NY 12571 USA
Phone: 845-758-0400
Fax: 845-758-2633
Email: curran@proceedings.com
Web: www.proceedings.com

TABLE OF CONTENTS

SPEAKER PRESENTATIONS

SESSION 1: ROLE OF VENTURING IN THE CHEMICAL INDUSTRY

SESSION 2: INCUBATION SPACE FOR CHEMICAL STARTUPS

SESSION 3: CHEMICAL STARTUP STRATEGIES

SESSION 4: LESSONS LEARNED AND STRUCTURING DEALS FOR SUCCESS

SESSION 5: OUTCOMES AND EXITS

SESSION 6: STRATEGIC VS. FINANCIAL INVESTING

STARTUP PRESENTATIONS

PART I

NCET2 STARTUP PRESENTATIONS

PART III

SABIC VENTURES
HELPING INNOVATIVE COMPANIES

WHO WE ARE

1

SABIC AT-A-GLANCE

1976 Company established	33,000 Employees around the world	50 Countries of operations	3rd Largest global chemical company*	120th Largest public company in the world*	3.96 US$ bn Estimated Brand Value**
85 US$ bn Total assets	5.7 US$ bn Net income	45 US$ bn Annual revenue	≈ 150 New products each year	11,738 Global patent filings	64 World-class plants worldwide

*Forbes 2018 **Brand Finance, 2019

2

OUR STORY: A HISTORY OF CHEMISTRY

1976

SABIC established

1980
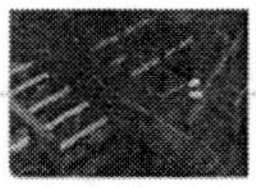
Hadeed founded

1980

JVs with Shell and ExxonMobil

1981

JV with Mitsubishi

1983

First products exported out of KSA

1984

First Public Shares

1996

Middles East's largest Joint Stock company

1997

New Logo introduced

2000

Selling in 100 Countries

2002

New Global Headquarters

2002

Acquired DSM Petrochemicals

2006

Acquired Huntsman Petrochemicals UK

2007

Acquired GE Plastics

2009

JV with Sinopec in China

2012

Inauguration of SABIC Academy

2015

JV with SK Chemical in South Korea

2016

Inauguration of Home of Innovation™

2017

Gulf Coast Growth Ventures JV with ExxonMobil

2018

Acquired 24.99% of stake in Clariant AG

2019
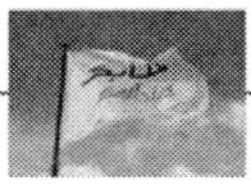
Brand Value estimated at $3.96bn by Brand Finance*

2019

Saudi Aramco agrees to acquire a 70% stake in SABIC

*According to 2019 report "Brand Finance Chemicals 10"

3

SABIC VENTURES

THE BUSINESS DEVELOPMENT CYCLE

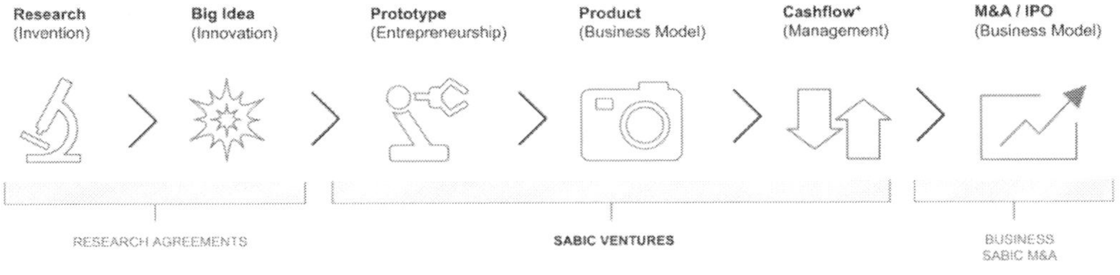

TECHNOLOGIES OF INTEREST

Alternative energy

Advanced materials, functional polymers and composites

Alternative feedstocks

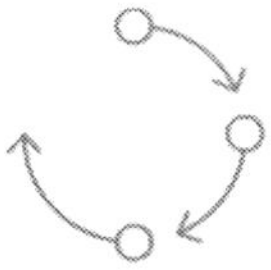

Process, resources and energy efficiency

OUR BUSINESSES

BUSINESS PORTFOLIO

PETROCHEMICALS

- Polyolefins
- PVC , Polyester and Polysterene
- Synthetic Rubbers
- Specialty Polymers & Polymer Additives
- Olefins, Oxygenates and Aromatics
- Chemical Intermediates & Industrial Gases

SPECIALTIES

- Engineered Thermoplastics
- Specialty Compounds
- Sheet & Film
- Composites
- Additive Manufacturing
- Fluids
- Thermosets and Additives

AGRI-NUTRIENTS

Nitrogen
- Prilled Urea
- Granular Urea
- Ammonia

Phosphate
- DAP and Dark DAP
- MAP

Specialty
- NPK
- TGU

METALS*

Long Steel
- Rebar
- Wire Rod
- Rebar in Coil

Flat Steel
- Hot Rolled Coils
- Cold Rolled Coils
- Galvanized Rolled Coils
- Pre-painted Rolled Coils

* Supplied under SABIC brand through Hadeed, a fully-owned SABIC Affiliate

STRATEGIC INDUSTRIES WE SERVE

AGRI-NUTRIENTS

Our portfolio of high quality agri-nutrients play a vital role in helping to secure adequate global food supply and make it possible to produce more food from less land.

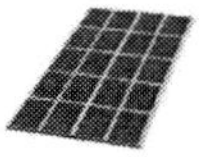

CLEAN ENERGY

SABIC's specialty and polymer products can help create sustainable renewable-energy solutions for the future.

ELECTRICAL & ELECTRONICS

Manufacturers need to make products in a cost-effective manner, yet meet consumers' expectations for style, ease of use and sustainability. Our materials are the foundation for the next generation of electronic devices.

CONSTRUCTION

We provide products to architects and builders who seek solutions to satisfy public demand for environmentally responsible structures while providing design freedom.

MEDICAL DEVICES

We help customers successfully address the challenges facing the industry by providing advanced technologies, a broad product portfolio, and a healthcare policy.

PACKAGING

SABIC polymers seek to help food and drink producers meet their goals while still providing sustainable benefits like lightweighting and environmentally-focused initiatives.

TRANSPORTATION

Automobiles, aircraft, rail interiors, heavy-duty trucks and other vehicles require new approaches to meet ever-rising requirements. We put our material, design and technical expertise to work to help OEMs meet this challenge and drive innovation to new levels.

8

TAILORED SUPPORT

WE WANT ENTREPRENEURS TO GROW AND WE ARE COMMITTED TO HELPING THEIR FUTURE BUSINESS SUCCEED

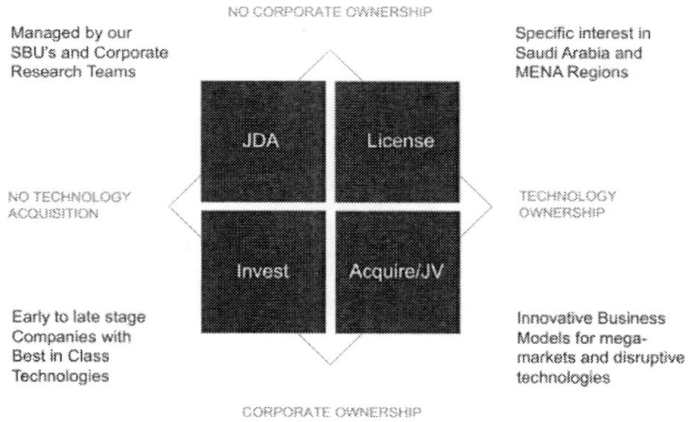

9

OUR PROCESS

Proposals reviewed for compatibility. A typical process goes through the following steps:

STRATEGIC FIT
MANAGEMENT TEAM
VALUE CREATED

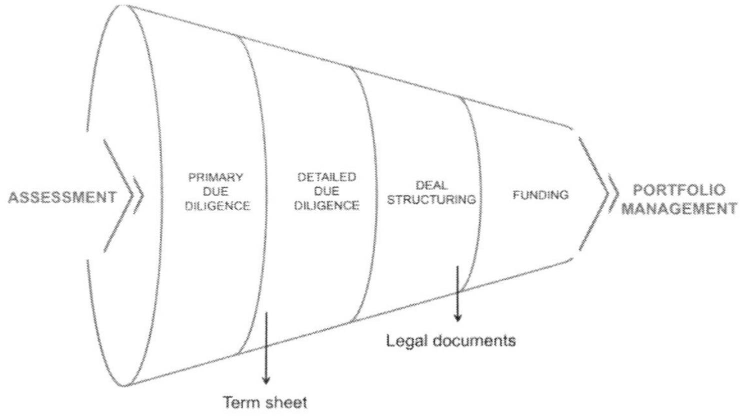

10

A PARTNER FOR GROWTH

We're looking for shared opportunities where we can build value for SABIC and others by putting these skills to work:

Creating value for our partners
Deep understanding of customer markets
Multi-market global reach
Scientific and technical expertise
Expertise in Intellectual Property
Large-scale R&D and product testing
Strategic Marketing & Planning
Saudi heritage
Access to local and global insight

11

سابك
sabic

SUBMIT A PROPOSAL

To submit a proposal please go to: ventures.sabic.com/submit-a-proposal

For more information: Ventures.sabic.com

CHEMISTRY THAT MATTERS™

12

سابك
sabic

DISCLAIMER

PLEASE NOTE THAT, UNLESS AND UNTIL WE ENTER INTO A WRITTEN AGREEMENT TO THE CONTRARY, ALL SUBMISSIONS ARE MADE ON A STRICTLY NON-CONFIDENTIAL BASIS. UNLESS APPROPRIATE ARRANGEMENTS HAVE BEEN AGREED WITH US IN ADVANCE, DO NOT SUBMIT INFORMATION THAT YOU CONSIDER TO BE CONFIDENTIAL OR PROPRIETARY.

The information presented in this presentation is provided solely for informational purposes, and is not intended to constitute an offer to enter into any type of partnership or business arrangement. Nothing presented herein is intended, or should be viewed, as an offer from SABIC Ventures, Saudi Basic Industries Corporation or any of their respective affiliates (hereinafter collectively referred to as "SABIC") to provide funding, or any type of support (financial or otherwise) to any individual, entity or business. SABIC makes no representation, warranty or guaranty, express or implied, concerning (i) the information contained herein (including as to accuracy, completeness or currency), or (ii) the type of business that might be considered a viable candidate for a shared opportunity. The information presented in this presentation is subject to change at any time, and SABIC has no duty to provide notice of any such change. Unless, and to the extent, otherwise expressly provided in a legally-binding agreement entered into by SABIC and the submitting party prior to the time of submission, all information submitted to SABIC, in response to any solicitation or statement set forth herein or otherwise, shall be submitted on a STRICTLY NON-CONFIDENTIAL BASIS. SABIC will not be responsible or liable for any loss, cost, damage or expense, whether direct, indirect, consequential or otherwise, including, without limitation, any claim for lost profits or lost opportunities, arising, or claimed to arise, from (i) any such submission of information, or (ii) reliance by any party upon any information or statement contained herein.

13

Air Liquide
creative oxygen

Air Liquide Venture Capital - ALIAD

THIS DOCUMENT IS PUBLIC
Wilmington, DE • April 23 2019
Air Liquide • ALIAD

Air Liquide - what we do

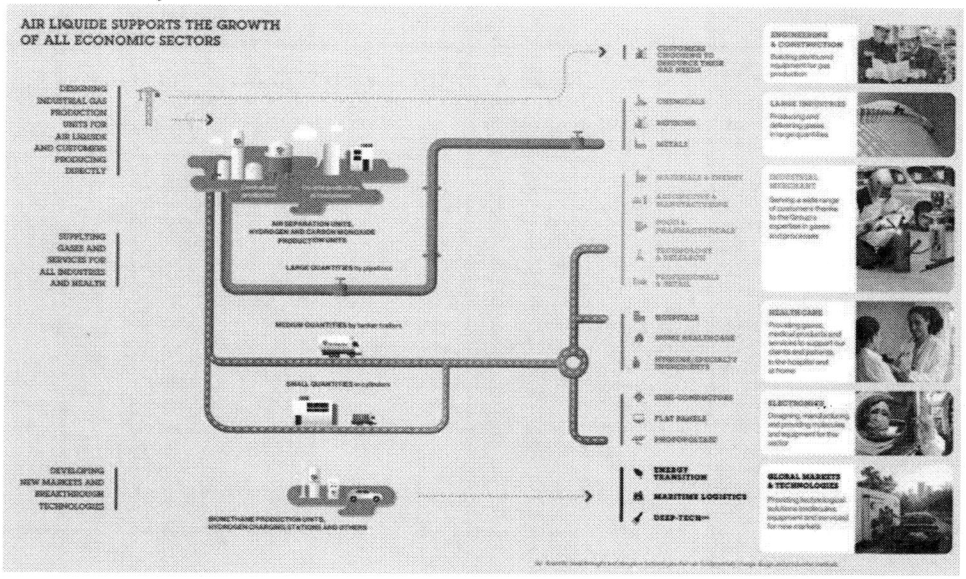

Air Liquide 2018

- Present in

80 COUNTRIES

~66,000 employees

A world leader in gas, technologies and services for Industry and Health

2018 GROUP REVENUE BY ACTIVITY

96% Gas & Services

27% Large Industries
44% Industrial Merchant
17% Healthcare
8% Electronics
2% Engineering & Construction
2% Global Markets & Technologies

27% of revenue for Gas & Services for Industry [a] realized in developing economies

Air Liquide by Geography 2018

EUROPE

33% Large Industries
30% Industrial Merchant
35% Healthcare
2% Electronics

AMERICAS

17% Large Industries
68% Industrial Merchant
10% Healthcare
5% Electronics

ASIA-PACIFIC

37% Large Industries
30% Industrial Merchant
4% Healthcare
29% Electronics

MIDDLE EAST & AFRICA

52% Large Industries
41% Industrial Merchant
7% Healthcare

Air Liquide innovation key figures 2018

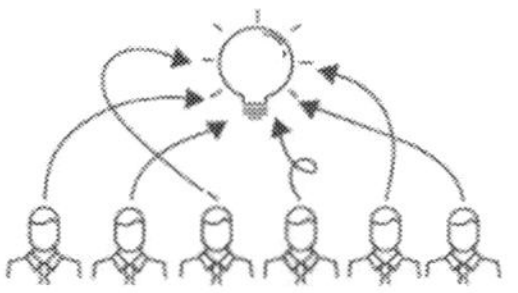

4,000

Employees contributing to innovation

€300 M

Innovation expenses

309

new patents

Air Liquide Venture Capital - ALIAD

Investing in technology companies of the future

- Supports growth of innovative technology start-ups
- Targets three sectors related to important social change:
  - **Energy transition**
  - **Healthcare**
  - **Digital**
- ALIAD is a strategic investor

Solidia Technologies

inpria®

Poly-Shape
3D Generative manufacturing

37
Investments since our creation in 2013

Total commitment of
~90 million euros

Inpria - EUV Photoresist

Inpria Tin Oxide Resists: Purpose Built for EUVL

5X smaller building blocks

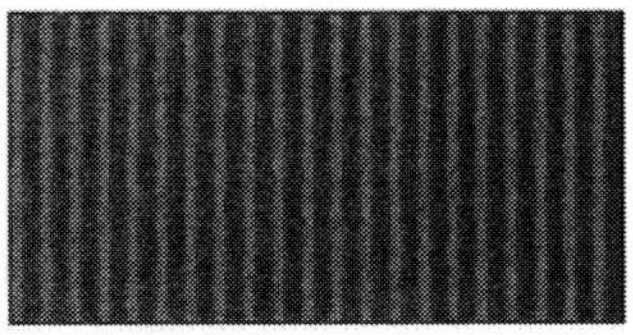

11nm half-pitch (Inpria)

Required resolution & smooth lines

5X more photons absorbed / volume

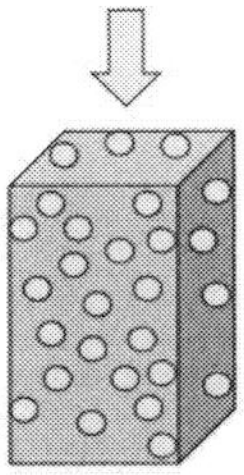

Faster photospeed at high resolution

10X higher etch selectivity

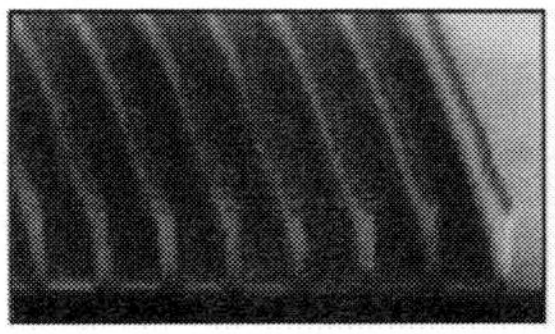

Improved yield & simpler process

Solidia Technologies

Solidia Cement^{TM} is cured with CO2

Solidia Technologies' patented system offers a profitable, high-performance path to the future of building materials.

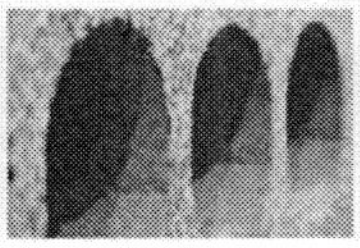

Thank You

CHAIN REACTION INNOVATIONS

Entrepreneurship at Argonne

MOVING
HIGH-IMPACT
SCIENCE INTO
HIGH-IMPACT
PRODUCTS

THE NEW ENERGY CRISIS IS CLIMATE CHANGE

1 More scientists to embrace the challenge
2 A better way to connect stakeholders to respond and adapt to the changing energy landscape
3 A better, faster, cheaper way to bridge the "valley of death"

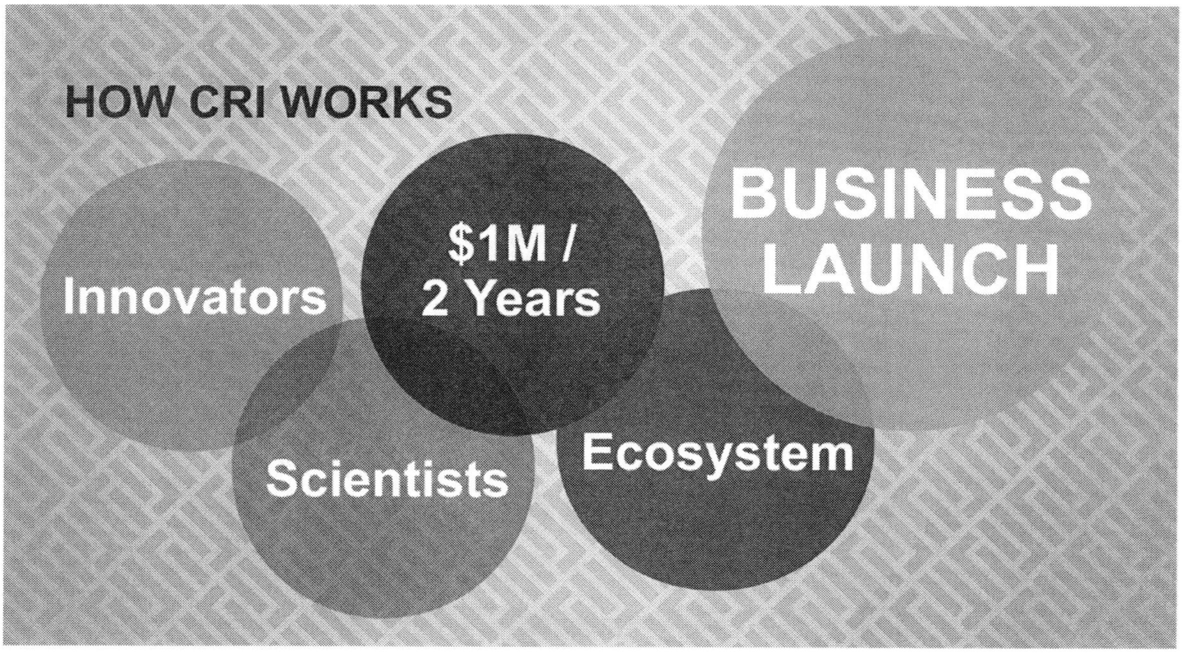
HOW CRI WORKS
Innovators
$1M /
2 Years
BUSINESS
LAUNCH
Scientists
Ecosystem

Collaboration between innovators and scientists
is the value generation engine for CRI

COHORT 1 OF CRI 2017-2019

Ian Hamilton

ATLAS ENERGY SYSTEMS, LLC

Felipe Gomez del Campo

fgc plasma SOLUTIONS

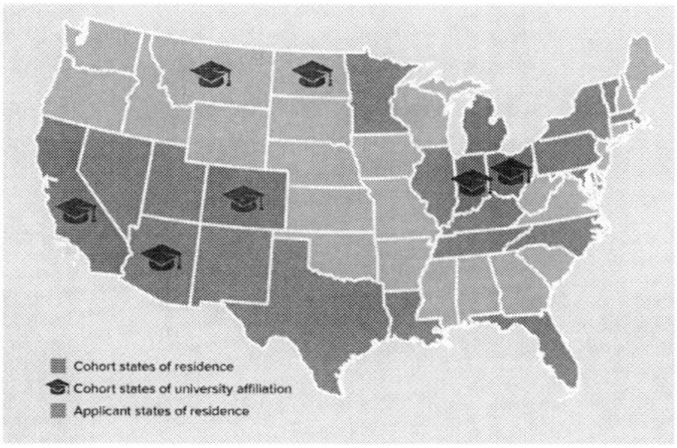

Tyler Huggins

Justin Whiteley

BTRFy

Julie Blumreiter

Bernard Johnson

ClearFlame ENGINES

U.S. DEPARTMENT OF ENERGY — Argonne National Laboratory is a U.S. Department of Energy laboratory managed by UChicago Argonne, LLC.

Argonne NATIONAL LABORATORY

U.S. DEPARTMENT OF ENERGY — Argonne National Laboratory is a U.S. Department of Energy laboratory managed by UChicago Argonne, LLC.

COHORT 2 OF CRI 2018-2020

Ted Seo

Volexion

Erika Boeing

Accelerate Wind

Chad Husko

iris

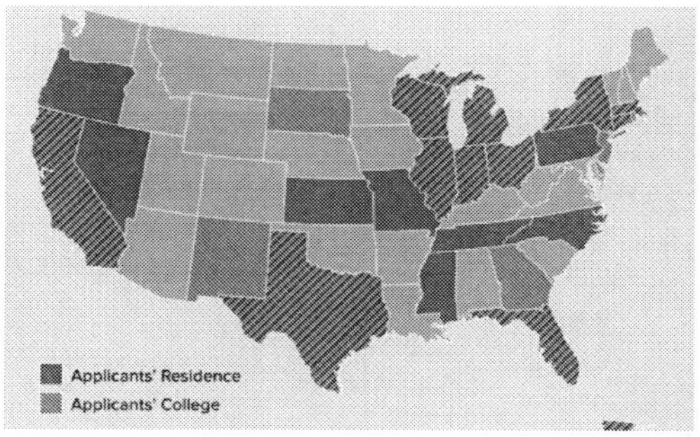

Tom Guarr

Veronika Stelmakh

MESODYNE

David Manosalvas-Kjono

AEROMUTABLE

ACTIVE RESPONSE TO AERODYNAMIC CHANGES

U.S. DEPARTMENT OF ENERGY Argonne National Laboratory is a U.S. Department of Energy laboratory managed by UChicago Argonne, LLC.

Argonne NATIONAL LABORATORY

COHORT 3 OF CRI 2019-2021

Khalid Alam
Freeze-Dried Biosensors for Water Quality and the Energy-Water Nexus

Yu Kambe
Direct Optical Lithography of Functional Inorganic Nanomaterials

Katie Kollhoff
Water Treatment Sorbents

Kevin O'Connor
Dielectric Materials for High Density Capacitive Energy Storage

Gary Ong
Nanocomposites for Elevated Temperature Hydrogen Technologies

U.S. DEPARTMENT OF ENERGY Argonne National Laboratory is a U.S. Department of Energy laboratory managed by UChicago Argonne, LLC.

Argonne NATIONAL LABORATORY

A new bridge between
discovery and market
adoption for energy startups
Argonne
NATIONAL LABORATORY
CHAIN REACTION
INNOVATIONS
Entrepreneurship at Argonne
chainreaction.anl.gov

UNIVERSITY OF DELAWARE
HORN ENTREPRENEURSHIP

Incubating (Chemical) Ventures:
Overview & Resources at UD

Christina Pellicane
Director of Commercialization Programs
Horn Entrepreneurship, University of Delaware

April 23, 2019
Built by entrepreneurs to power a better world.

University of Delaware (UD)

- $1.36B endowment (71st globally)
- 80+ research centers and institutes
- 7 colleges & 3 schools
- Land-, sea- and space-grant institution
- #8 Chemical Engineering program
- 5 strategic pillars:
  - Enhancing the success of our students
  - Building an environment of inclusive excellence
  - Investing in our intellectual and physical capital
  - Strengthening interdisciplinary and global programs
  - **Fostering a spirit of innovation and entrepreneurship**

University of Delaware (UD) est. 1743

Innovative Human Capital and Research

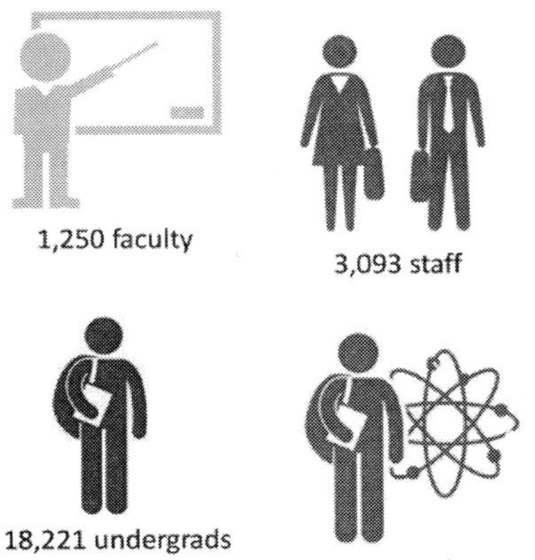

1,250 faculty

3,093 staff

18,221 undergrads

4,164 grad

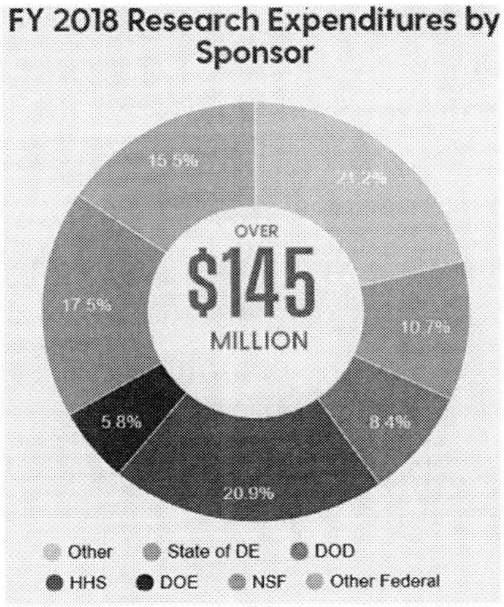

Since 2009:
465 Invention Disclosures
130+ patents issued
58 licenses

UNIVERSITY OF DELAWARE
HORN ENTREPRENEURSHIP

UD Innovation & Entrepreneurship Ecosystem

UNIVERSITY OF DELAWARE
HORN ENTREPRENEURSHIP

DELAWARE TECHNOLOGY PARK
A State, University & Private Sector Partnership

UNIVERSITY OF DELAWARE
ECONOMIC INNOVATION & PARTNERSHIPS

Incubation Space, Connections, Resources

DTP@STAR INCUBATOR

AMERICA'S SBDC DELAWARE

DELAWARE BIOTECHNOLOGY INSTITUTE®

DELAWARE INNOVATION SPACE

UNIVERSITY OF DELAWARE
HORN ENTREPRENEURSHIP

Horn Entrepreneurship Overview

Horn Entrepreneurship (est. 2012) utilizes Evidence-Based Entrepreneurship methodology & is a resource for **everyone at UD** via:

- **Academic programs**: major, minors, certificates, and electives
- **New Venture Development programs**: grants, training, and mentorship for all faculty, staff, post-docs and students
- **Events**: Open to the community for professional development, enrichment, and networking

UNIVERSITY OF DELAWARE
HORN ENTREPRENEURSHIP

Office of Economic Innovation & Partnerships

- OEIP (est. 2008) is responsible for commercialization of inventions based on UD's Intellectual Property
- Functionalities within OIEP:
  - Technology Transfer
  - Small Business Development Center
  - Corporate Partnerships
  - Procurement Technical Assistance Center

UNIVERSITY OF DELAWARE
ECONOMIC INNOVATION
& PARTNERSHIPS

AMERICA'S SBDC DELAWARE

UNIVERSITY OF DELAWARE
HORN ENTREPRENEURSHIP

Delaware Technology Park

- DTP (est. 1992) has housed more than 75 companies & helped create more than 16,000 new jobs in the community
- $300M invested in DTP and an additional $300M given to DTP organizations through research grants
- Provides wet lab & office space
- New satellite facility at STAR campus

DELAWARE TECHNOLOGY PARK
A State, University & Private Sector Partnership

UNIVERSITY OF DELAWARE
HORN ENTREPRENEURSHIP

Commercialization Myth

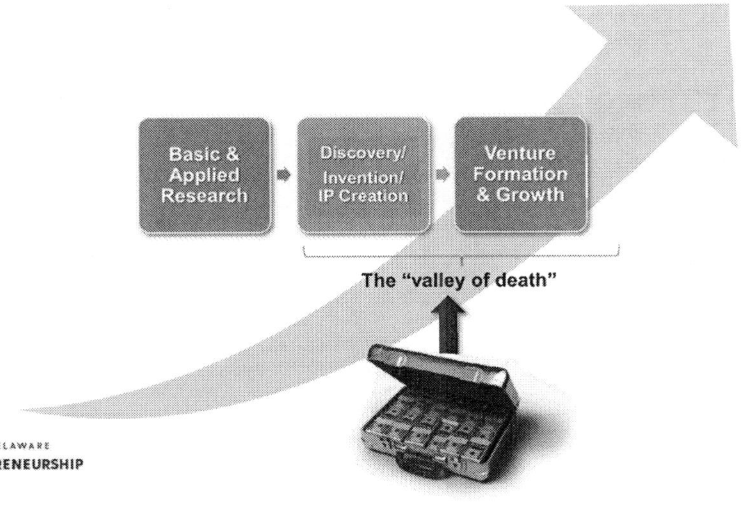

UNIVERSITY OF DELAWARE
HORN ENTREPRENEURSHIP

Commercialization Reality

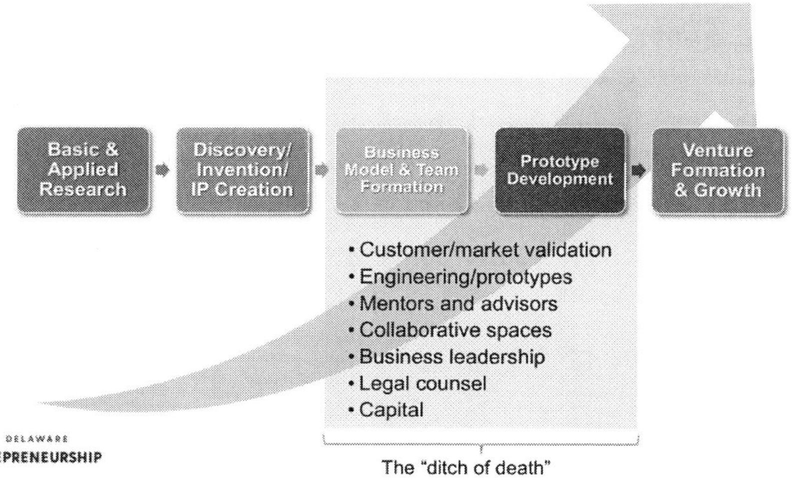

UNIVERSITY OF DELAWARE
HORN ENTREPRENEURSHIP

New Venture Development Programs

Startup Village
Guidance from Industry Professionals

VentureOn
Mentoring, Support & Resources

Venture Mentoring Service
Mentoring for Startups

Tech Innovation Showcase
Celebrating UD Startups & Inventors

UNIVERSITY OF DELAWARE
HORN ENTREPRENEURSHIP

New Venture Development Programs, cont.

NSF I-Corps Site
Grants for STEM Technologies

Summer Founders
12-Week Pre-Accelerator

Blue Hen Proof of Concept
Funding & Training

Hen Hatch
Business Concept Funding Competition

UNIVERSITY OF DELAWARE
HORN ENTREPRENEURSHIP

New Venture Development Programs, cont.

NSF I-Corps Site
Grants for STEM Technologies

Funding: NSF grants of $3,000 are given to each selected team and can be used to fund expenses related to customer discovery research, including solution-testing costs

Training: Four sessions introducing lean startup methodology and customer discovery research

Resources: Access to the Venture Development Center's co-working space 24/7 and access to VentureOn

Follow-on Opportunities: Eligible for NSF I-Corps Teams and UD Blue Hen Proof of Concept Program

UNIVERSITY OF DELAWARE
HORN ENTREPRENEURSHIP

New Venture Development Programs, cont.

Summer Founders
12-Week Pre-Accelerator

Funding: Students (undergrad and grad) receive a stipend & customer discovery funding while working on their ventures fulltime

Training: 12-week pre-accelerator program

Resources: Includes weekly mentoring and education sessions in addition to progress meetings with an advisor

Demo Day: Culmination of lessons learned in front of investors, partners, etc.

UNIVERSITY OF DELAWARE
HORN ENTREPRENEURSHIP

New Venture Development Programs, cont.

Blue Hen Proof of Concept
Funding & Training

Funding: $75k grants (with UD IP) & $10k grants (no UD IP) & $75k postdoc fellowship

Training: Expected to pursue NSF I-Corps Teams ($50k grant) and SBIR/STTR grant ($225k)

Resources: Monthly BH-POC team meetings, regular mentor meetings, quarterly investor/partner lunches

UNIVERSITY OF DELAWARE
HORN ENTREPRENEURSHIP

New Venture Development Programs, cont.

Hen Hatch
Annual Business Concept Funding Competition

Eligibility: Entrepreneurial students, alumni, faculty and staff

Goals: Make connections, collect feedback on ideas and compete for startup cash and prizes totaling $100,000

This Year: Tuesday, April 30th 5:00pm to 9:00pm (The Queen, Wilmington)

RSVP: www.udel.edu/henhatch

UNIVERSITY OF DELAWARE
HORN ENTREPRENEURSHIP

New Venture Development: Impact

- Engaged 253 alumni and community members in venture support programming; provided **1,099+ hours of mentoring and advisory services**
- $815k in funding to UD startups awarded by Horn Entrepreneurship
- Actively supporting 60+ startup ventures led by UD students, postdocs and faculty. Collective accomplishments include:
  - **6 ventures placed in top 5 in national startup competitions**, including:
    - 1st place - HBCU pitch competition
    - 3rd place - Southeastern Hedge Fund competition
    - 1st place - DOE Launch-R competition
  - 15 ventures won **more than $1.7M in follow on funding** (Jul. 1 – Dec. 31, 2018)
  - 20+ SBIR/STTR proposals submitted (most currently pending)
  - Ventures have reported contracts, revenues and other **receivables > $6M**
- Won **NSF phase II award** to continue UD/Horn's I-Corps Site program

UNIVERSITY OF DELAWARE
HORN ENTREPRENEURSHIP

Success Stories

NUVVE

V2G Technology
Inventor: Willett Kempton, College of Earth, Ocean, & Environment

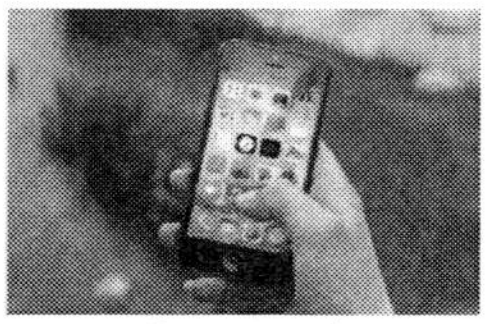

FingerWorks
Inventors: Wayne Westerman and John Elias, College of Engineering

Reebok

STF Technologies
Inventor: Norm Wagner, College of Engineering

GenPro
Genome Profiling, LLC
Accelerating the Promise of Precision Health

Founder: Adam Marsh, College of Earth, Ocean, & Environment

BASF
We create chemistry

Biofungicide technology
Inventors: Janine Sherrier & Harsh Bais, College of Agriculture

GO BaBY go!
mobility & sociability

Founder: Cole Galloway, College of Health Sciences

UNIVERSITY OF DELAWARE
HORN ENTREPRENEURSHIP

Success Stories (in the making)

Healthcare & Biotech

- Avkin
- CatX
- ImmunoStrip Test
- Resonate Forward
- Cancer Dormancy
- Veni Diagnostics

Energy & Cleantech

- SHI
- RjKarbon
- Lectrolyst
- W7energy

Agriculture

- SAGE SMART GARDEN
- Miscanthus Propagation
- Carbon Sorbents
- Grass Sterility

Advanced Materials

- SMARTenius
- MCET Tech

Electronics & Robotics

- TiRiC
- TFETLogic

Software

- GeoSwap
- Lyapunov Technologies
- CYBER 20/20

UNIVERSITY OF DELAWARE
HORN ENTREPRENEURSHIP

Delaware I & E Ecosystem - 2012

DE MEP
Delaware Manufacturing Extension Partnership

SCORE Delaware

SBA
U.S. Small Business Administration

Delaware
Division of Small Business

Delaware Sustainable Chemistry Alliance

First State Innovation

University of Delaware
Horn Entrepreneurship

Emerging Enterprise Center
A Place to Launch and Grow

Delaware I & E Ecosystem - 2019

Siegfried

DE MEP
Delaware Manufacturing Extension Partnership

Capital One

SCORE Delaware

TS

Delaware Prosperity Partnership

nfte

Pete du Pont
Freedom Foundation

SBA
U.S. Small Business Administration

{Coded by: Kids}

Next Fab

DECA

Delaware
Division of Small Business

Delaware Sustainable Chemistry Alliance

Delaware IQ

First State Innovation

Schoolyard Ventures

NIN
National Innovation Network

Great Dames

The Mill

Dual School

STRIDE
Science, Technology and Research Institute of Delaware

NCC innovates
Office of Economic Development

Delaware Youth Leadership Network

Technology Forum of Delaware

1313 Innovation

DelawareBio
Advancing Bioscience Innovation

Ye!

University of Delaware
Horn Entrepreneurship

Startup Island

DFi
Delaware Founders Initiative

Emerging Enterprise Center
A Place to Launch and Grow

Thank you

- Meet some of our tech startups at our VentureConnect tables!
- Email: cpell@udel.edu
- Website: udel.edu/horn
- Venture Development Center: 132 East Delaware Avenue, Newark, DE

UNIVERSITY OF DELAWARE
HORN ENTREPRENEURSHIP

Commercialization Pathway

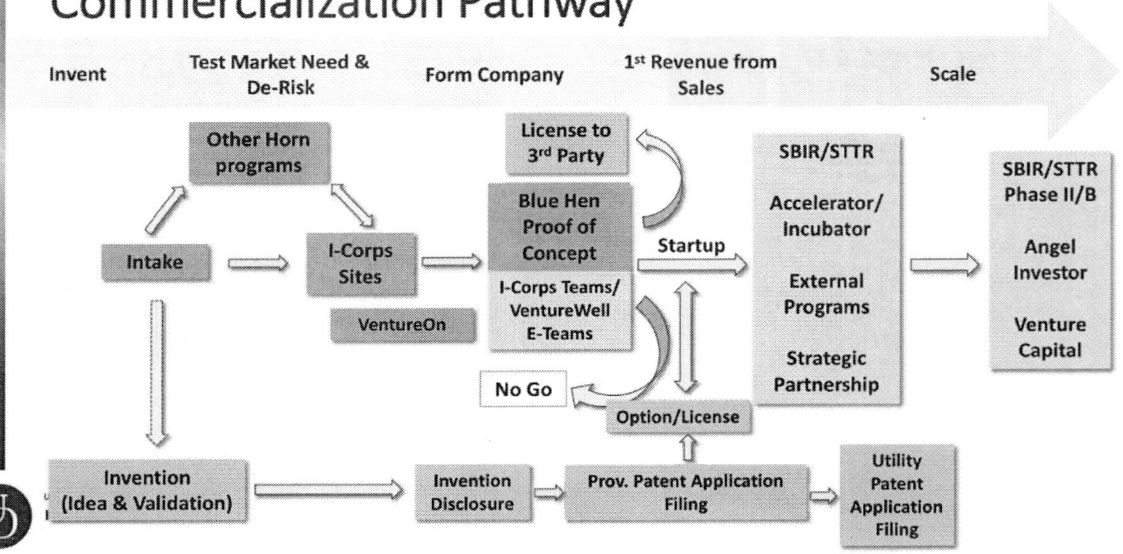

Commercialization Goals

- Increase the impact of UD investment by **closing the gap** between research and initial developmental and pre-commercialization efforts
- Grow UD's innovation **community** by engaging students, faculty and external stakeholders (philanthropists, investors, entrepreneurs, corporations, government)
- Evaluate commercial potential, demonstrate the value of technology (for SBIR/STTR/I-Corps) and generally **de-risk technologies** to investors (for startups) and commercial partners (for licensing)

UNIVERSITY OF DELAWARE
HORN ENTREPRENEURSHIP

Success Metrics

- Entrepreneurial Mindset
  - Increased human innovation capital
  - Innovation skills necessary to compete in today's workforce
- Institutional Culture Change
  - Shifting from research as knowledge creation to societal impact
  - Support P&T policies, startup-friendly licensing deals, programs developed to support faculty commercialization, etc.
  - Creation and growth of the entrepreneurial community
- Startup Outcomes
  - Jobs created, IP licenses, funding raised, revenue generated

UNIVERSITY OF DELAWARE
HORN ENTREPRENEURSHIP

Key Takeaways

- Reasons innovators are motivated:
  - Opportunity to make a significant societal and economic **impact** with their research (licensing or startup route)
  - Gaining **innovation skills** necessary to compete in today's workforce
- Goal is to de-risk commercial potential for **startup** OR **licensing** route
- Offer **training** programs/workshops & **connections** to high-quality **mentors**, subject matter **experts,** and business **services**
- Offer **funding** for meeting with potential customers and building prototypes and/or proving the technical concept
- It takes a **village** to support (chemical) deep tech ventures

UNIVERSITY OF DELAWARE
HORN ENTREPRENEURSHIP

CORPORATE-STARTUP DEVELOPMENT PROGRAM

NCET2 National Council of Entrepreneurial Tech Transfer

TONY STANCO
Executive Director, NCET2
tony@ncet2.org

How we do it

- DISCOVERY
- DEVELOPMENT
- FUNDING

NCET2 MISSION:
Discover, develop, and fund corporate-aligned startups from universities, angel groups, and state organizations through the Corporate-Startup Development Program specifically aligning startups to corporate business units

Corporate-Startup Development Program

DISCOVERY

DISCOVERING corporate-aligned startups for NCET2 Fortune 500 Corporate Members

- Using Corporate Needs Solicitations (wish lists) that describe the corporate's business needs and technology interest areas
- From university startups, angel portfolios, and state funding organizations
- We package, review, and present startups through Virtual Startup Pitches to corporates for corporate investment and strategic partnership

DEVELOPMENT

DEVELOPING startups with Fortune 500 corporate interest through NCET2 Startup Development Officers (SDOs)

- Usually startups are too early so development is needed to align to corporates
- SDOs develop startups with milestones from corporates to guide startup development
- Aligning startups to corporate exits and strategic partnerships
- SDOs are former Fortune 500 open innovation and venture executives, active angel investors, serial entrepreneurs, startup attorneys with extensive knowledge and experience in startup development

FUNDING

FUNDING corporate-aligned startups through the National Angels Group and our Investment Banker Fund

- Specifically created to invest in corporate-aligned startups that NCET2 develops through the Corporate-Startup Development Program

NCET2 National Council of Entrepreneurial Tech Transfer

Virtual Startup Pitches

It's a way to discover startups of online startup pitches to Fortune 500 corporates

- Begins on June 5, 2019 at 1:00pm ET/10:00am PT (regularly scheduled sessions thereafter)
- An easy and convenient platform to watch startup pitches at your desk
- We show 5-10 startups at a time per showcase in 4 main technical areas, Pharma and Medical Devices, Advanced Manufacturing, Chemicals, Clean Tech, Information and Communications Technology
- Startup Development Officers (SDOs) provide feedback to startups
- Startup submission deadline May 17, 2019

DISCOVERY DEVELOPMENT FUNDING

NCET2 National Council of Entrepreneurial Tech Transfer

Startup Development Officers

50

STARTUP DEVELOPMENT OFFICERS (and growing)

Who are the SDOs?

- Former Fortune 500 Venture and Open Innovation Executives
- Serial Entrepreneurs
- Angel Investors
- Startup Attorneys

What do they do?

- Manage corporate development for startups
- Part of startup Management Team as a corporate business development and early exit advisor
- Develop startups to line up for corporate exits
- Startup general counsel services

DISCOVERY DEVELOPMENT FUNDING

NCET2 National Council of Entrepreneurial Tech Transfer

Startup Development Officers

50

STARTUP DEVELOPMENT OFFICERS (and growing)

- Extensive knowledge and experience in startup development
- Support startup corporate business development activities
- Are advisers to startups and startup C-Level executives to line up startup for corporate exits
- Guides startup through corporate-recommended milestones
- Guidance to the startups for:
    - IP development
    - Science team development
    - Management team development
    - Milestone development
    - Product development
    - Sales projections
    - Gross margins
    - Price/earnings gross ups
    - Total global potential market
    - Exit price
    - General counsel services

DISCOVERY DEVELOPMENT FUNDING

NCET2 National Council of Entrepreneurial Tech Transfer

Corporate-Startup Development Program

40	250	700
STARTUPS IN DEVELOPMENT	STARTUPS IN DISCOVERY	STARTUPS SUBMITTED SINCE 2016

DISCOVERY DEVELOPMENT FUNDING

NCET2 National Council of Entrepreneurial Tech Transfer

How Can You Engage?

Corporates: Be an NCET2 Corporate Member

- Join the Virtual Startup Pitches starting on June 5, 2019
- Register by emailing RiaAncheta@ncet2.org

Startups: Submit your startups for the Virtual Startup Pitches for corporate review and SDO feedback

- Send non-confidential pitch decks to submissions@ncet2.org
- Deadline: May 17, 2019

Universities and Angel Groups: Sign up for Corporate Needs Solicitations

- Email: RiaAncheta@ncet2.org

DISCOVERY DEVELOPMENT FUNDING

NCET2 National Council of Entrepreneurial Tech Transfer

CORPORATE-STARTUP DEVELOPMENT PROGRAM

NCET2 National Council of Entrepreneurial Tech Transfer

TONY STANCO
Executive Director, NCET2
tony@ncet2.org

DISCOVERY DEVELOPMENT FUNDING

NCET2 National Council of Entrepreneurial Tech Transfer

Chemical Ventures in the University Setting

David Mackanic

AIChE Chemical Ventures Conference, April 24th, 2019

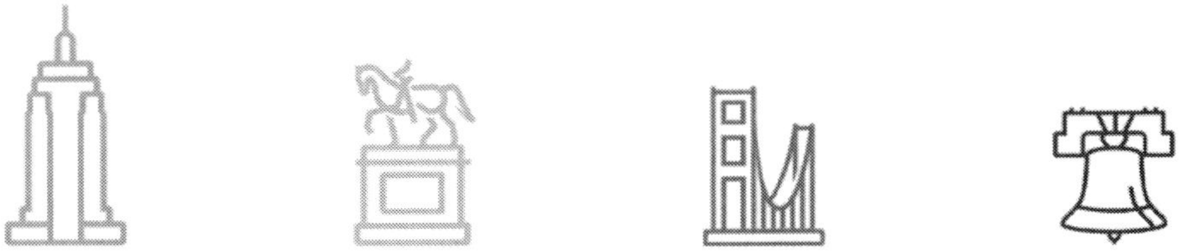

Introduction – David Mackanic & Dorm Room Fund

David Mackanic

Partner, Dorm Room Fund

Ph.D. Candidate, Chemical Engineering, Stanford University

B.S. Mechanical Engineering, B.A. Chemistry, Virginia Tech

Interests: Materials, energy storage, polymers, hardware

Dorm Room Fund

First student-run VC in the country (est. 2012)

Offices in SF, NYC, Boston, and Philadelphia

Sponsored by **First Round Capital**

2

Outline of Presentation

1. Introduction to Dorm Room Fund

2. The case for investing in University-based startups

3. Portfolio highlights – DRF investments in the chemical space

4. Case studies of high-potential University-based startups

5. Conclusion

3

Outline of Presentation

1. Introduction to Dorm Room Fund

2. The case for investing in University-based startups

3. Portfolio highlights – DRF investments in the chemical space

4. Case studies of high-potential University-based startups

5. Conclusion

4

36 Partners Across 19 Schools

With Dorm Room Fund partners across undergraduate and graduate programs around the country, we have 36 active partners across Undergraduate, Master's, MBA, and PhD programs across the country.

Boston College · Brown · Columbia University in the City of New York · Columbia Business School · Johnson Cornell University · Princeton University · Harvard University · Harvard Business School · Massachusetts Institute of Technology · MIT Management Sloan School · Northeastern · NYU · NYU Stern · Stanford University · Stanford Graduate School of Business · Berkeley University of California · BerkeleyHaas · Penn University of Pennsylvania · Wharton University of Pennsylvania · Yale

5

Portfolio Performance

151 Investments

Between our very first investment in Firefly (acquired in 2014) and our latest investment earlier this week, we've written over 150 checks.

Over the last 5 years, we've been funding the strongest community of student entrepreneurs, consisting of founders across Undergraduate, MBA, Master's, and PhD programs around the country.

6

$300M+ in Follow–On Capital

Our portfolio companies have gone on to raise over $300M in funding from investors like Sequoia Capital, Andreessen Horowitz, Y-Combinator, and more.

We're starting to see traditional VC's invest with more confidence and frequency in student-founded startups than ever before.

BESSEMER VENTURE PARTNERS

G/

USV Union Square Ventures

ANDREESSEN HOROWITZ

Y

HIGHLAND CAPITAL PARTNERS

SEQUOIA

First Round

BainCapital VENTURES

7

Successful Exits

ZODIAC
Customer analytics
(Acq. by Nike)

QuickHelp
Edtech

Firefly
Web services

A&B AMERICAN STYLE
BBQ Sauce

preo
Hospitality tech

SKILLBRIDGE
Freelance finance consulting platform

tbh
Social Media
(Acq. by Facebook)

Py
Hiring

Distributed systems
Cryptocurrency
(Acq. by Coinbase)

8

Student Founders Are Everywhere

9

Outline of Presentation

1. Introduction to Dorm Room Fund

2. The case for investing in University-based startups

3. Portfolio highlights – DRF investments in the chemical space

4. Case studies of high-potential University-based startups

5. Conclusion

10

Why invest in Universities?

Conventional Wisdom: Universities are innovation centers → high presence of startup companies

Universities encourage high-tech plant openings

Do Universities lead to more high-tech plant openings?

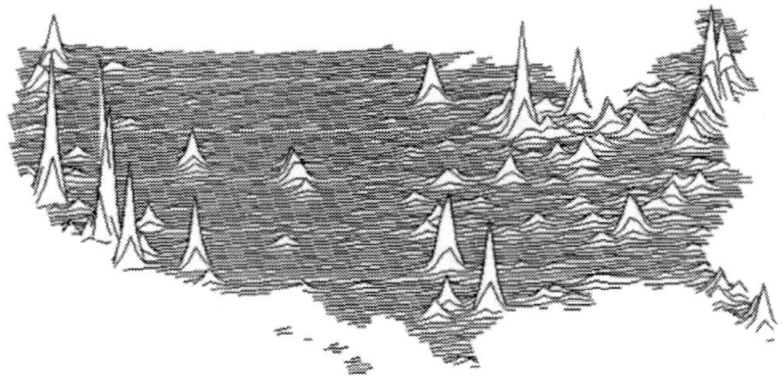

⊥University Research and Development spending within 60 miles of plant
**Significant at 5% level
***Significant at 1% level

While effect is less important than direct economic levers, University R&D leads to an increase in high-tech plant openings.

Data from U.S. Census Bureau and Bureau of Labor Statistics, 1997-2000

(Guimaraes, et al., *Beyond the Silicon Valley: University R&D and high-technology location,* Journal of Urban Economics, 60, 2006)

12

Universities encourage high-tech ***chemical*** plant openings

Results within the chemical industry (inorganic, organic, agricultural, misc.)

Variable	Dirichlet-Multinomial regression result	Significance Level
Labor Costs	-0.396	**
Taxes	-0.033	**
Weighted University R&D⊥	0.0021	**
Sample size	12,008	--

⊥University Research and Development spending within 60 miles of plant
**Significant at 5% level
***Significant at 1% level

The opening of high-technology chemical plants is influenced by the presence of University R&D within 60 miles of the plant.

Data from U.S. Census Bureau and Bureau of Labor Statistics, 1997-2000

(Guimaraes, et al., *Beyond the Silicon Valley: University R&D and high-technology location*, Journal of Urban Economics, 60, 2006)

13

Outline of Presentation

1. Introduction to Dorm Room Fund
2. The case for investing in University-based startups
3. Portfolio highlights – DRF investments in the chemical space
   - OneThree Biotech
   - CleanCult
   - Nextbiotics
4. Case studies of high-potential University-based startups
5. Conclusion

14

OneThree Biotech: Using Data to Accelerate Drug Discovery

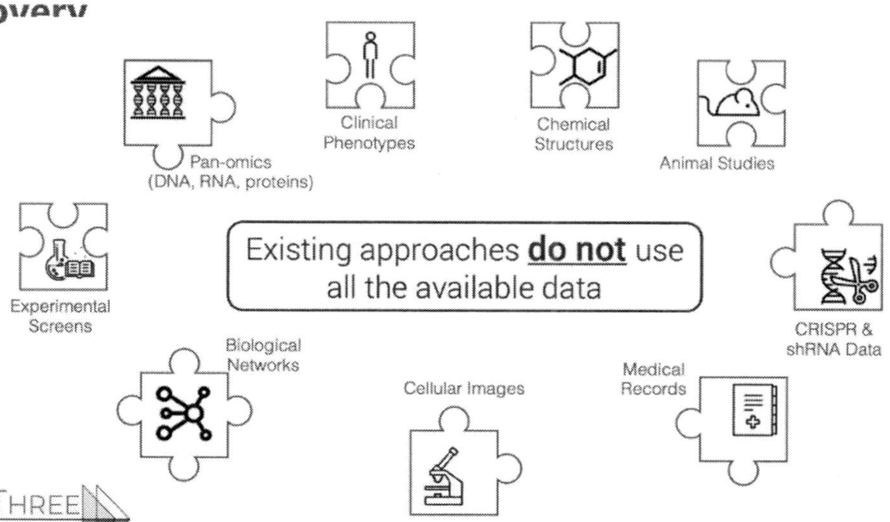

ONETHREE biotech

15

OneThree Biotech: Using Data to Accelerate Drug Discovery

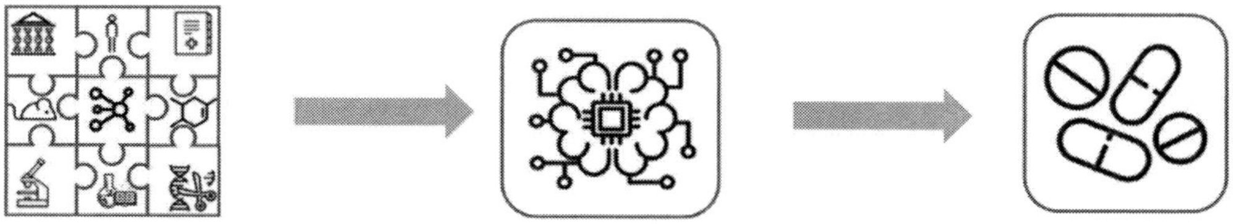

Data Integration	End-to-End AI	New Treatments
27 data types: 10M drug candidates; 3K diseases,	Insights across multiple steps with a 90% accuracy	Identify and derisk new targets and treatments

ONETHREE biotech

OneThree model accurately predicted outcomes of **15 drugs** currently in clinical trials

16

OneThree Biotech: The IP Challenge

Founding Team is a Cornell PhD Student and a Cornell Professor

Neel Madhukar, PhD
Founder, CEO
PhD -- Computational Biology, Forbes "30under30", Experience working in experimental bio, AI, and tech

Olivier Elemento, PhD
Founder
Director, Institute for Precision Medicine, Weill Cornell Medicine
Leading expert in AI drug dev

Challenge: Intellectual Property

- Technology developed and patented at Cornell
- Licensing deal trades a small stake in the company and minor revenue sharing in exchange for an exclusive license
  - Cornell also agreed to allow continued research and development on-campus, forming a long term technology partnership with shared licensing

Cornell University

ONETHREE biotech

University relationship enables ongoing development of deeply scientific, functional technology in a low-risk setting.

17

Nextbiotics: Engineering Viruses for Good

Nextbiotics is developing a biotechnology platform that uses proprietary algorithms and synthetic biology to engineer viruses to destroy pathogenic bacteria.

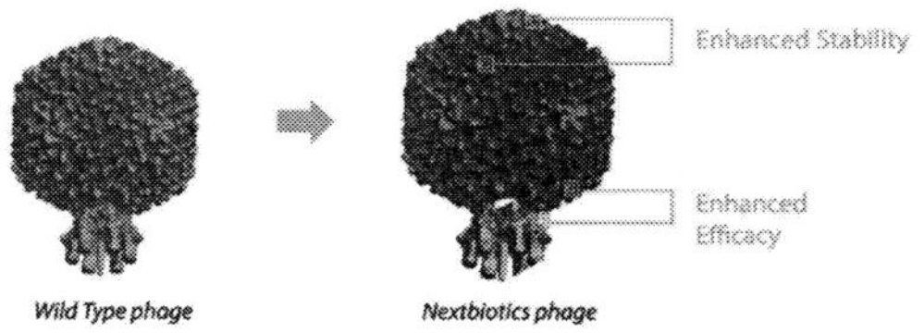

NEXTBIOTICS

18

Nextbiotics: Engineering Viruses for Good

Founding Team:
Jorge– UC Berkeley PhD advised by Jennifer Doudna, world-leading CRISPR expert
Maricel- Former consultant with operational experience

↓

Spin-out Process
Rapidly iterated on target market for their technology through the Berkeley Skydeck program.
Leveraged university resources to acquire interns, kept all science/IP development outside of University

↓

Current position
Partnering with a major pharma player to continue to develop technology and identify market opportunities

NEXTBIOTICS

Berkeley SKYDECK

DRF

World-expert mentors and access to University funding positioned Nextbiotics for success

19

CleanCult: Direct to Consumer Chemicals

Puerto Rico Science, Technology & Research Trust

Partnered with former Puerto Rican pharmaceutical manufacturers to produce natural emulsifiers from coconuts

Challenge: How to upsell this commodity chemical product?

cleancult

20

CleanCult: Creating a Brand

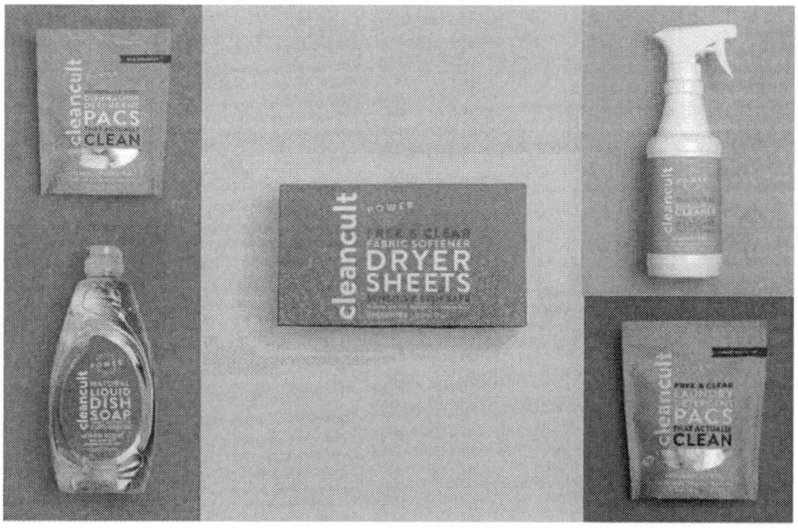

21

CleanCult: Creating a Brand

Sells monthly plastic-free refill packages via a subscription model

cleancult

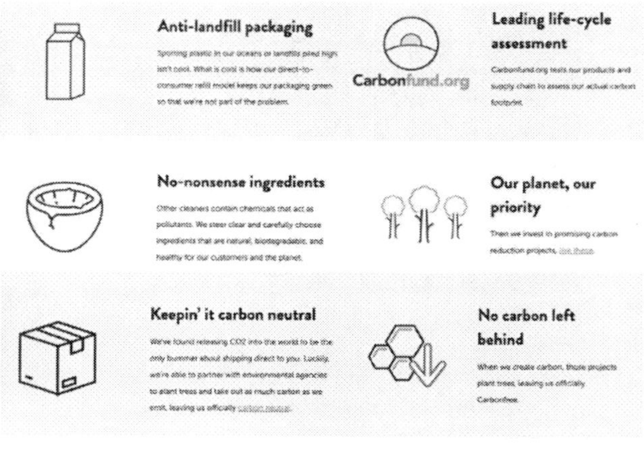

Identifying shifting consumer behavior to D2C, environmentally conscious products enables growth of Cleancult brand

22

Outline of Presentation

1. Introduction to Dorm Room Fund
2. The case for investing in University-based startups
3. Portfolio highlights – DRF investments in the chemical space
4. Case studies of high-potential University-based startups
   - Antora Energy
   - Lilac Solutions
5. Conclusion

23

Antora Energy: Thermal Energy Storage

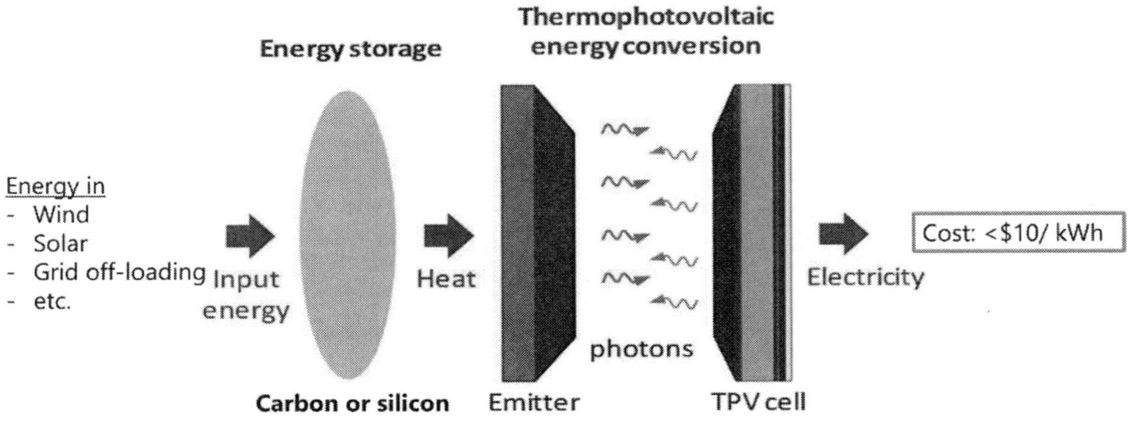

Antora Energy

(Datas et al., *All silicon energy storage*, 31st European Photovoltaic Solar Energy Conference and Exhibition, 113, 2014)

24

Antora Energy: Utilizing University Resources

Initial Team:
Two Stanford Engineering students

→

TomKatCenter FOR SUSTAINABLE ENERGY STANFORD UNIVERSITY

Initial Funding
TomKat Center at Stanford – Seed grant and interns
Order ~ $10Ks

→

cyclotronroad

Spin Out Process
Cyclotron Road Accelerator @ Lawrence Berkeley
Order ~ $100Ks

→

arpa-e CHANGING WHAT'S POSSIBLE

R&D Budget
ARPA-E grant order ~ $1Ms

→

Today
Technology de-risked via extensive R&D. Raising funding round right now to grow team and expand business focus

Antora Energy successfully utilized University resources to provide a long runway and prove-out technology

Antora Energy

25

Lilac Solutions: Identifying a lithium need

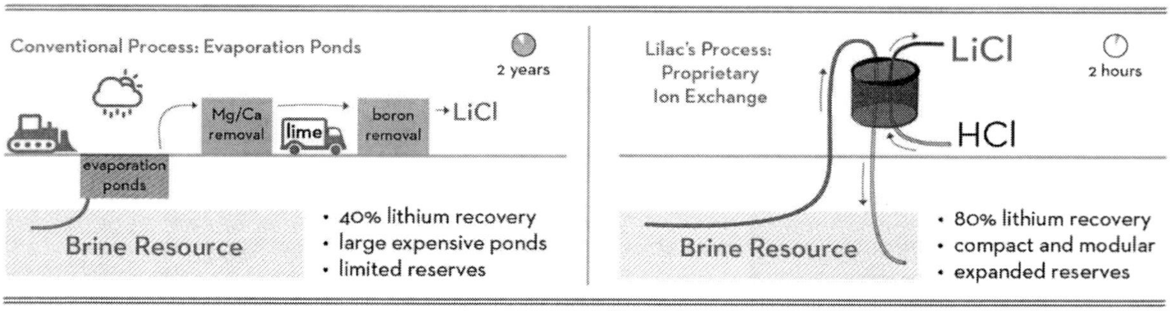

Improve IRRs for lithium extraction projects from 15% to 75%

Lilac solutions

26

Lilac Solutions: Responsible Fundraising

Northwestern University →

Founder: ·
PhD Student @ Northwestern
Focused on computational methods
No IP

SBIR America's Seed Fund →

Pre-Seed Funding
SBIR focused on computational development of materials

Seed Funding:
Angels/VCs; focused on building out physical processes
Moved to Berkeley to share lab space and hire engineers

→ **Pre-A**
Led by institutional investors in the energy space. Focused on proving out pilot plant capability.

→ **Today**
- 8 + Patents filed
- Pilot plant operational
- raising series A

Lilac solutions raised the right amount of money as their technology progressed, significantly de-risking the investment opportunity

Lilac solutions

27

Outline of Presentation

1. Introduction to Dorm Room Fund
2. The case for investing in University-based startups
3. Portfolio highlights – DRF investments in the chemical space
4. Case studies of high-potential University-based startups
5. Conclusion

28

In conclusion, Universities are a vibrant source of chemical-focused startups

Dorm Room Fund investment strategy in chemical space

1. Major breakthrough in a growing field (*OneThree*)
2. Thoughtful development of intellectual property and defensibility (*OneThree, Nextbiotics*)
3. Work with world-class leaders in a particular field (*Nextbiotics, OneThree*)
4. Identify a opportunity to capture shifting consumer behavior (*Cleacult*)

General "best practices" for University-based chemical startups

1. Leverage University resources for initial resources and sustainable growth opportunities
2. Raise funds responsibly, adequately de-risking tech development as company grows

Questions?

ONETHREE biotech

cleancult

NEXTBIOTICS

DRF

AIChE Management Division

AIChE young PROFESSIONALS

Dorm Room Fund – investment strategy in chemical space

Two trends we have invested in

- Computational/data enabled chemical discovery (bio company)
- Direct to consumer brand plays to distribute chemicals (Cleancult)

Case Study: How to "win" as a University-Based Startup

- Case study Antora Energy
- Case study – Lilac Solutions

30

Developing and Launching New Bio-Based Businesses in the Chemical Industry

Presented to the Chemical Ventures Conference
April 24, 2019

Ray W. Miller
Verdecute Consulting
"helping you execute sustainable growth"

The Sorona® Journey

1946 - first DuPont internal report on PTT performance
1960's, 70's, and 80's – worked off and on towards various chemical PDO options
1993 – Nylon Enterprise Transformation Team formed; Degussa develops PDO route
1994 - Shell announced Corterra® PTT to replace nylon in carpet
2Q95 – first isolated bio-PDO; Genencor collaboration began;
1Q96 – first batch in Carother's autoclave – using Degussa chemical PDO
3Q96 – first batch made in a plant autoclave
4Q96 – first PTT carpet fiber spun in Seaford, DE
2Q97 – **first DuPont commercial PTT (batch) production**, broad bio patent allowed
4Q98 – Chemical PDO plant opened in Wesseling, Germany
2Q99 – Chemical PDO source qualified, Tate & Lyle selected for Bio
1Q00 – launched Sorona® with chemical PDO using a converted CP at Kinston, NC
2002 – achieved commercial Bio-PDO™ performance in pilot plant
2003 – DuPont spins off Nylon as Invista
2004 – Mohawk Smartstrand™ Carpet Launched
2005 – broke ground for Bio-PDO™ plant
2006 – 100 MM ppy Bio-PDO™ plant started up – first shipment
2009 – FTC granted new "Triexta" generic; Shell exited PTT
2009 - 2017 Major growth of Sorona® in carpet and fabric markets
2010 – DT&L announced 35% Bio-PDO™ expansion
2018 – DT&L announced another 30% Bio-PDO™ expansion

Chemical 5 years

Bio 10 years

Verdezyne Platform Development Timeline

verdezyne

Creating Unique Solutions For a Sustainable Future

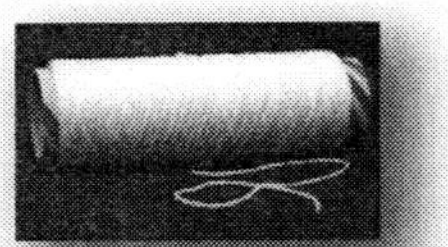

2008
- Founded in 2008
- Private Biotech
- > 40 Patents
- > 60 Employees
- Carlsbad, CA

2010
- POC for Adipic Acid and DDDA – Industrial, Materials, Specialty Chemicals, Lubricant, Fragrances

2012
- World's first biobased N6,6 fiber
- Pilot Scale; ADA and DDDA
- Sold cellulosic ethanol technology to DuPont

2014
- Commercial Partners appointed

CONNELL BROS.

will&co

2016
- POC for carotenoids – F&B, Animal Feed, Nutraceuticals, Personal Care

2017
- Successful launch and sales of FerroShield with tolling agreement – Metalworking, Coolant

2018
- May 15, 2018 ceased further platform development
- VerdePalm plant was on track for completion in 2018
- Verdezyne Assets and IP for sale (see Rynetech Bio)

MV | MONITOR VENTURES

OVP VENTURE PARTNERS

bp

DSM

Sime Darby

verdezyne

3

<u>Business Development *Issues* in *Larger* Companies</u>

1. **Existing customers control resource allocation.**
2. **Small new markets don't solve the growth needs of large companies.**
3. **The ultimate value of disruptive technology is unknowable in advance.**

So the barriers to using disruptive technologies in existing markets by existing suppliers becomes the attacker's advantage by smaller companies, *if they capitalize quickly!*

Key Learnings

New-to-the-market bio products require developing value chain know how and value-in-use pricing; drop-ins are much easier but pricing is more competitive.

A credible threat to an existing business can be a more powerful motivator than the opportunity to form a new business. New entrants have the attacker's advantage.

Strategic partnerships really can make a difference.

Examples of Possible Bio-based Platforms

Oxygen-containing are favored

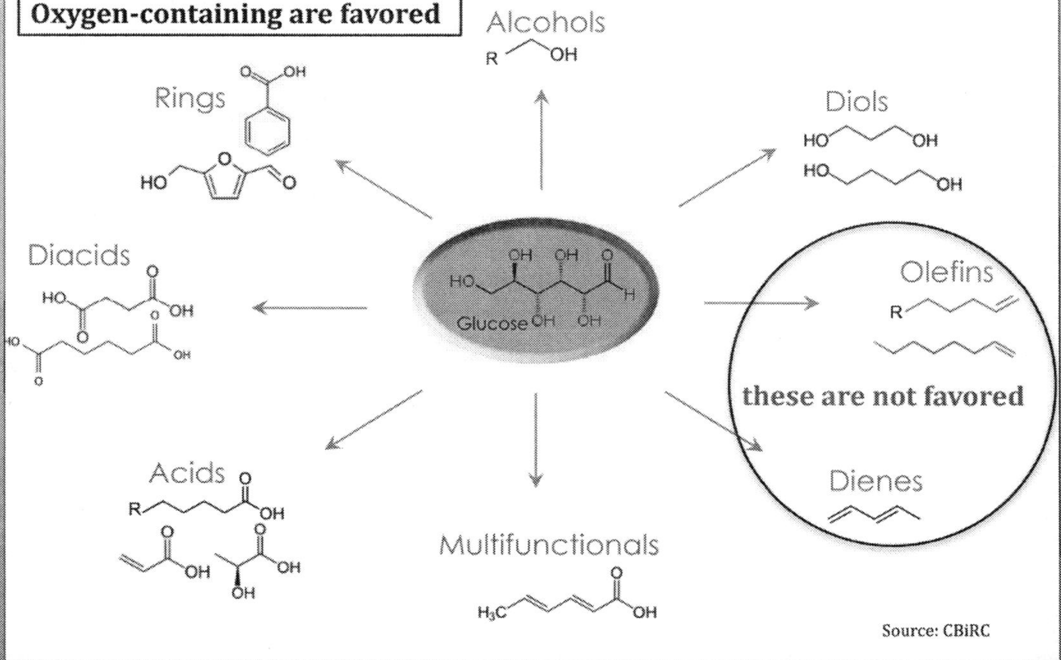

Source: CBiRC

Thank you
Verdecute Consulting, Inc.
helping you execute sustainable growth
Ray W. Miller
President
720 Ann Drive
Kennett Square, PA
19348-1552
1-302-530-1175
millerrayw@verizon.net

PANGAEA
VENTURES LTD.

ADVANCED MATERIALS
VENTURE CAPITAL

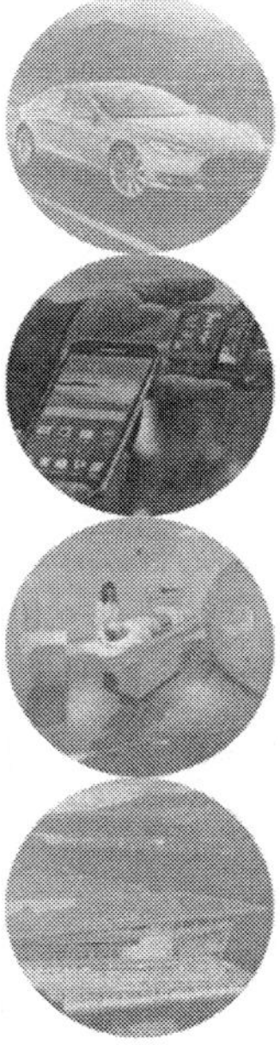

Introduction to
Pangaea Ventures

PANGAEA
VENTURES LTD.

The world leader in

Advanced Materials Venture Capital

- Materials-driven impact strategy
- Underserved market segment
- 25 world-leading strategic limited partners
- Offices in Canada and USA

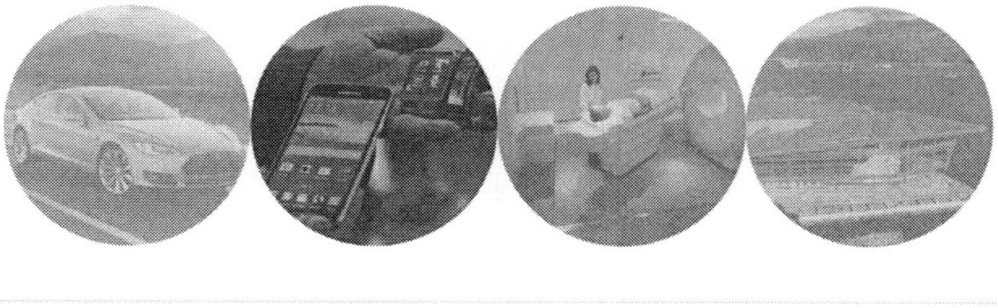

2

PANGAEA VENTURES LTD.

Advanced Materials to Make Our World Better

Energy	Electronics	Health	Sustainability
–Renewables –Energy Storage –Efficiency	–Semiconductors –Display –Robotics	–Medical devices –Nutrition –Regenerative med	–Green chemistry –Agriculture –Water

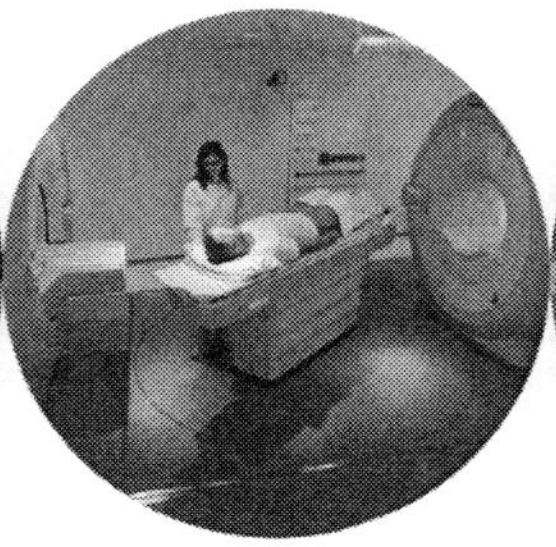

3

PANGAEA VENTURES LTD.

Pangaea's Material Impact

- Pangaea tracks and reports on its portfolios' cumulative impact
- Estimated cumulative portfolio impact by 2025:

58M
tons CO_2 reduction

39B
m^3 fresh water saved

7.7M
tons food production increased

10.7M
lives impacted

Global Challenges + Advanced Materials Innovation = Impact Opportunity

PANGAEA
VENTURES LTD.

Advanced Materials VC Experts

Chris Erickson
General Partner
Founder, 18 years VC

Purnesh Seegopaul
General Partner
PhD Chem, Deep Technical

Keith Gillard
General Partner
Corporate VC, AgTech

Andrew Haughian
General Partner
Energy, Scale-up

Janelle Goulard
Director, Health Investment
GE Health, Start-up

Matt Cohen
Director, Technology
Materials Start-up, Technical

Tracy Hedberg
Director, Finance & Admin
Audit, Finance & Reporting

Kookai Chaimahawong
Associate
Impact investing, UN

USA

Canada

5

PANGAEA
VENTURES LTD.

World-Leading Strategic LPs

- 25 Fortune 500 multinationals (sportswear company not shown)
- Unequalled network of materials leaders

AGC | AsahiKASEI | BASF We create chemistry | BEKAERT better together | Castrol | CoorsTek Amazing Solutions

DOOSAN | EVONIK INDUSTRIES | Henkel | JSR | Lam RESEARCH | Mitsubishi Chemical Holdings

muRata Innovator in Electronics | NITTO DENKO | GC | سابك sabic | SAMSUNG | SEKISUI

Severstal | Shin-Etsu | SOLVAY | TOSOH | TOYOTA TSUSHO CORPORATION | umicore

6

PANGAEA
VENTURES LTD.

The world leader in

Advanced Materials Venture Capital

- Materials-driven impact strategy
- Underserved market segment
- 25 world-leading strategic limited partners
- Offices in Canada and USA

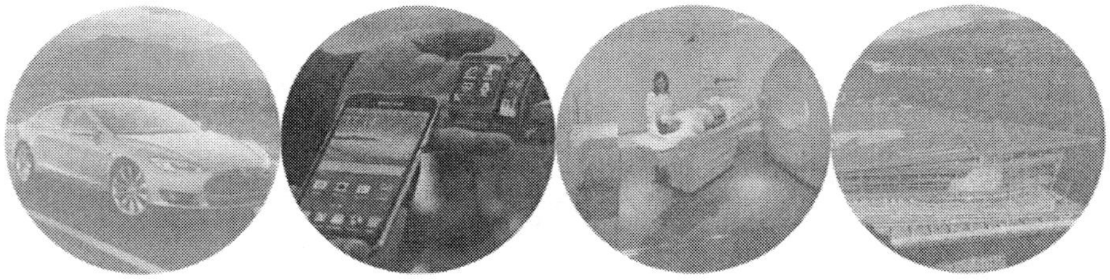

7

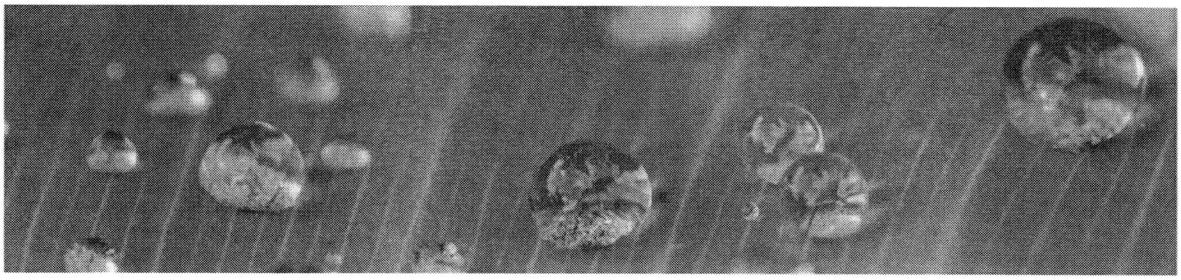

emerald
Technology Ventures

Emerald Technology Ventures Introduction for Investors

Connecting capital with industrial innovation

Global Leader in Industrial Technology Venture Capital

Emerald's key figures at a glance

64 Portfolio companies in Europe, North America and Israel	**>15%** Net IRR on recent funds
Investing in industrial technology innovation since **2000**	Million EUR under advisory from corporations, institutional investors & family offices **600**
1200 Business plans p.a.	**24** Corporate Investors

2 | Confidential

emerald
Technology Ventures

Our Approach

Three pillars to successfully deliver value to entrepreneurs and investors

Expertise	Experience	Relationships

Expertise in Fund and Mandate Management

Managing Industrial Technology Investments since 2000

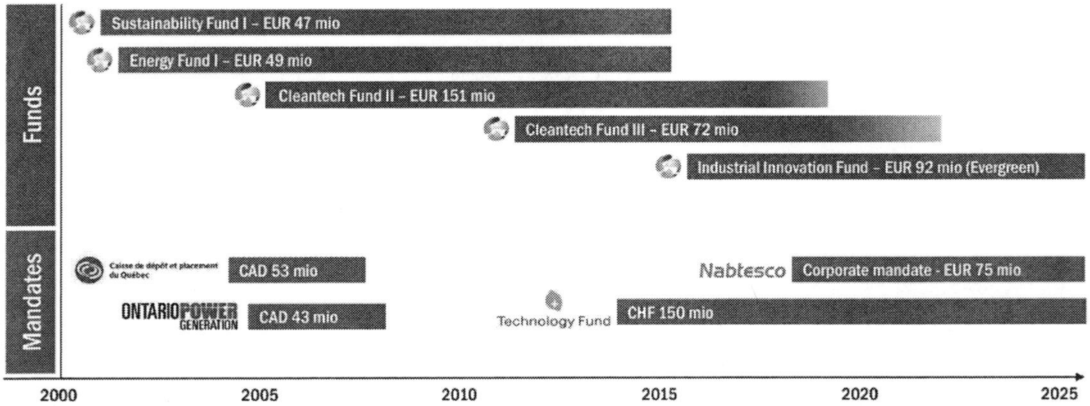

Expertise	Experience	Relationships

emerald Technology Ventures

Emerald's Professional Investment Team

Interdisciplinary team with the experience, expertise and capacity

emerald Technology Ventures

Stable Team with Deep & Diversified Expertise

Experience in industrial sectors, technology, corporate finance and venture capital

emerald Technology Ventures

Hundreds of Transactions in North America and Europe

64 portfolio investments linked to different technologies, markets, jurisdictions and cultures

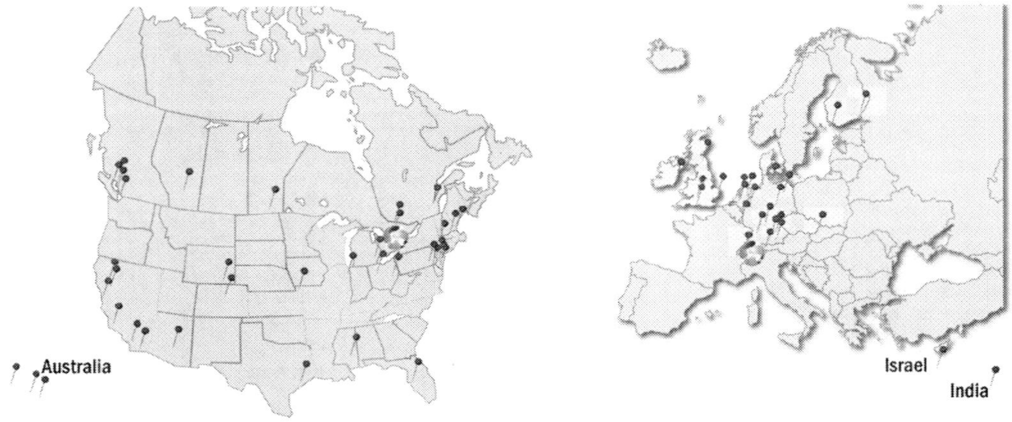

Expertise | Experience | Relationships

emerald Technology Ventures

150 Years of Combined Industrial Technology VC Experience

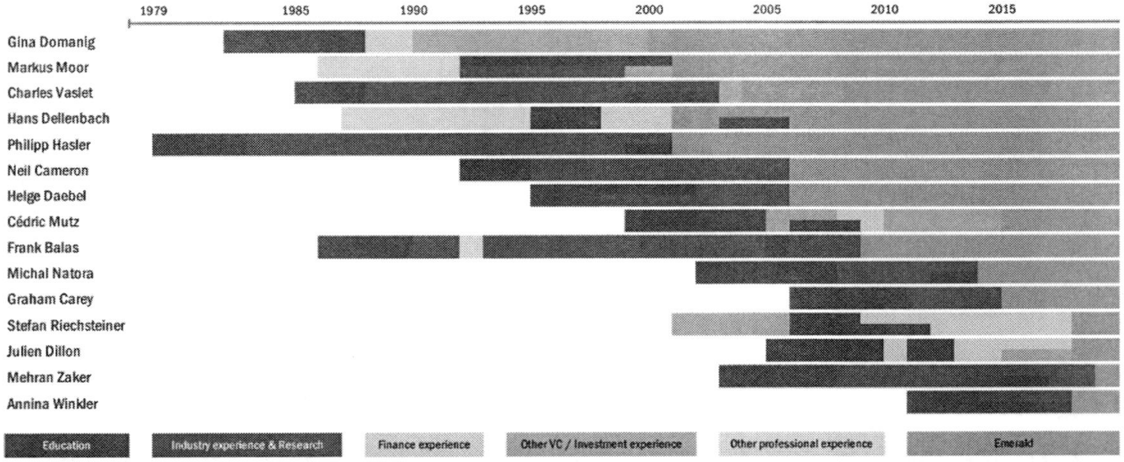

Expertise | Experience | Relationships

emerald Technology Ventures

Building Emerging Industry Leaders

Emerald portfolio companies making an impact

Smart Infrastructure	OT Cyber Security	Biogas	Advanced Materials
pure technologies — Emerald driven merger with PPIC, listed on TSX and subsequently acquired by Xylem	Security Matters — acquired by ForeScout	Schmack Biogas AG — listed on Deutsche Börse and subsequently acquired by Viessmann	PEMEAS Fuel Cell Technologies — acquired by BASF

Biopesticides	Water Treatment	Electric Drives	Smart Water
AgraQuest — acquired by Bayer CropSciences	inge — acquired by BASF	VISEDO Electricity in Motion — acquired by Danfoss	Optimatics PLAN SMARTER — acquired by Suez

Expertise | Experience | Relationships

emerald Technology Ventures

Global Network to Deliver Value

24 Corporate investors in current Emerald funds

ABB, BEKAERT, CAT, CLARIANT, COLRUYT GROUP, dic

DOOSAN, ECOLAB, EVONIK POWER TO CREATE, Henkel, Heraeus, IREON VENTURES

JSR, MAHLE, MICHELIN, Nabtesco, سابك sabic, SCG

SHV ENERGY, SIBELCO, SK, SULZER, SUNCOR ENERGY, WM WASTE MANAGEMENT

Expertise | Experience | Relationships

emerald Technology Ventures

Focus: Early & Expansion Stage Opportunities in N. America & Europe

...and outstanding teams that can create successful companies within five years

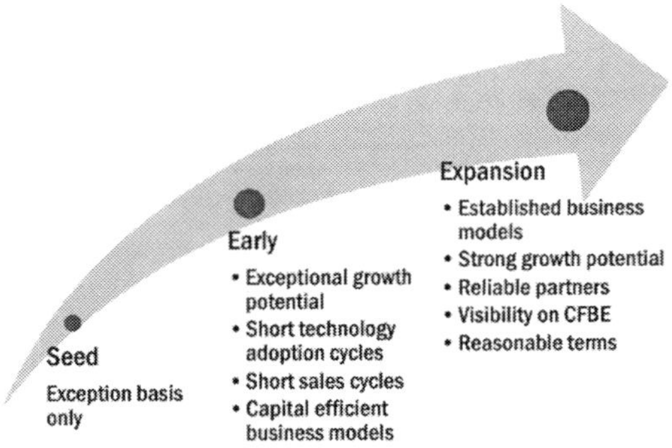

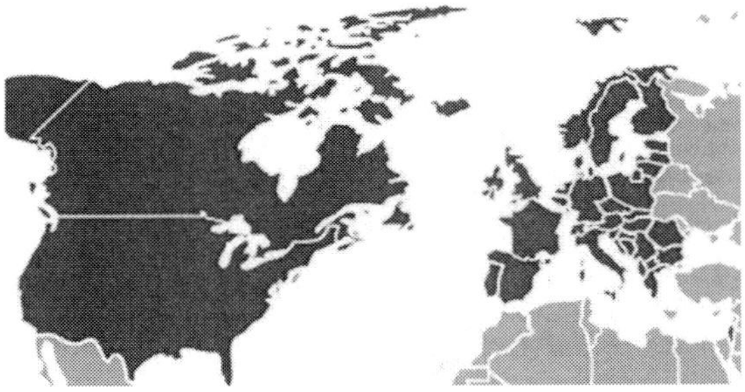

Sector	Stage	Geography

emerald Technology Ventures

Contact

www.emerald-ventures.com

European Office

Emerald Technology Ventures AG
Seefeldstrasse 215
8008 Zurich, Switzerland

T: +41 44 269 61 00
F: +41 44 269 61 01
info@emerald-ventures.com

North American Office

MaRS Centre, West Tower
661 University Ave., Suite 445
Toronto (ON) M5G 1M1, Canada

T: +1 416 900 3453
F: +1 416 900 3457
info@emerald-ventures.com

emerald Technology Ventures

SDA Ventures LLC

- Founded in 2011 by Principals Steve Dubin and David Abramson, former CEO and President, respectively, of Martek Biosciences.
- SDA Ventures LLC assists emerging growth companies through large multinational companies in the functional ingredient, nutritional products and health & wellness markets, on matters including corporate development, business acquisition, customer relations, growth strategies and corporate finance.

Sample Current and Past SDA Clients

PAINE SCHWARTZ PARTNERS TECHNOLOGY CROPS INTERNATIONAL Generation UCAN i-Health NOBLEGEN persona

SDA Ventures LLC

1

Steve Dubin, Principal SDA Ventures

- Served as the CEO of Martek Biosciences, from July 2006-2011.
- Joined Martek in 1992, initially as CFO and General Counsel.
- Chairman of Enzymotec Ltd, a leading global supplier of specialty lipid-based products and solutions technologies from January 2014 until its acquisition by Frutarom Industries in January 2018.
- Other experience includes working with a variety of early-stage technology companies and spending 12 years in various positions at Suburban Bank, now part of Bank of America, including Vice President and Treasurer of its venture capital subsidiary, Suburban Capital Corporation.

SDA Ventures LLC

2

Martek Biosciences

- Founded in 1985 by scientists from Martin Marietta while working with NASA to explore the use of microalgae as a food supply for long term space exploration.
- Developed into a microbial biotechnology company.
- Martek produced DHA and ARA used in infant formula and DHA for use in foods, beverages and supplements.
  - By 2011, over 64 million babies worldwide had been fed Martek ARA/DHA, included in 99% of infant formulas in the U.S.
  - Martek's DHA used in 350 food and beverage products around the world.
  - $300 million+ in infant formula ingredient sales annually by 2012, total company sales near $500 million.
- Acquired by Royal DSM NV for $1.09 billion in 2011.
  - 35% premium to Martek's closing share price.
  - 8.5 times EBITDA.

life's DHA™ HEALTHY BRAIN, EYES, HEART

life's ARA™ by MARTEK

SDA Ventures LLC

3

Strategic vs. Financial Investing

Charles (Chuck) Paul
Chemical Ventures Conference April 24, 2019

Henkel

Henkel

Henkel Ventures

2 Henkel

Who we are

Adhesive Technologies

LOCTITE
TECHNOMELT
TEROSON

Beauty Care

Schwarzkopf
Dial
syoss

Laundry & Home Care

Persil
all
Purex

Adhesive Technologies Five Business Areas

#1 Adhesive Company
Virtually every manufacturing industry

Packaging and Consumer Goods	Transport and Metal	General Industry	Electronics	Consumer, Craftsmen and Building
		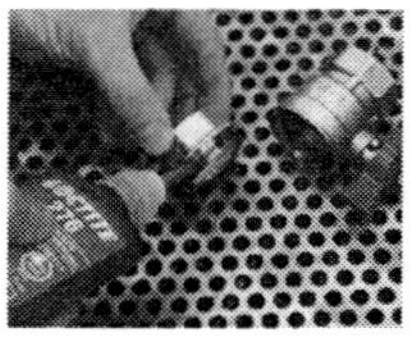		

Henkel Ventures
Search fields for Adhesive Technologies

1	3D Printing	"Materials and processes that stretch the envelope of additive manufacturing—from functional prototypes to large scale construction materials and general manufactured parts."
2	Printed Electronics	"Advanced electronics materials and technologies enabling flexible, printed electronics — from (medical) sensors, printed heaters to RFID antenna and flexible circuitry."
3	Batteries	"Electrochemical cells (in particular thin-film and high capacity), their active and passive thermal management, and mechanical protection."
4	Thermal Management	"Insulating or active heat management components with focus on electronics, insulation in construction and coatings for various industries."
5	Smart Building Materials	"Materials that enhance specifics of or replace traditional materials, and innovate production and assembling processes in building, construction and furniture."
6	Functional & Super-Hydrophobic Coatings	"Smart coatings that bring additional functionality on to a surface—from protective through medical, anti-microbial & anti-fingerprint to non-fluorinated super-hydrophobic coatings."
7	Smart Packaging	"Low cost printed electronics packaging applications to enhance consumer interaction—from RFID & NFC chips to printed sensors."
8	Sustainable Solutions & Concepts	"Solutions building on natural, environmentally friendly, health/wellbeing & safe concept."

Visit henkel-ventures.com for more information

5 Henkel

Collaborations by Search Field

3D Printing	10
Printed Electronics	8
Batteries	0
Thermal Management	5
Smart Building Materials	1
Functional Coatings	12
Smart Packaging	2
Sustainable Solutions	6

Other 17

May 19 6 Henkel

Our Current Portfolio

KRIYA MATERIALS

nano-metal oxides for coatings/films

DropWise Technologies Corp.

nano-coatings via i-CVD.

VITRIfLEX

flexible, transparent ultra-barrier films for electronics

nbd nano

nano-materials for durable repellency

Copprint

nano-copper conductive inks

Venture Fund Investments:

emerald Technology Ventures

PANGAEA VENTURES LTD.

CHINA MATERIALIA LLC

CIRCULARITY CAPITAL

7 Henkel

Partnering with Start-Ups

- Provide them
  - access to customers
  - access to supporting technology / testing
- Provide us
  - understanding of the opportunities
  - potential for growth

8 Henkel

Adjacencies – Acquire or Invest?

- Established / Understood by Henkel - M&A

- New / Dynamic / Poorly Understood by Henkel - CVC

9 Henkel

Good Partnering - Examples

- Product Safety and Regulatory Support for Global Supply Chain
- Introductions to Major Customers via Key Account Managers
- Credibility / Viability at Major Customers
- Application Specific Testing for Highly Specified Applications
- Complementary Primers and Sealants

Title of the presentation May 23, 2019 10 Henkel

Negative Partnering - Examples

- Too Slow – Choose Another Partner
- Exclusive JDA – Discourages Better Partners
- Insufficient Internal Pull – lost time and wasted efforts
- Poorly Researched Market Opportunity
- Pivot to Market of No Strategic Interest

Thank you!

SOFINNOVA
PARTNERS FOR LIFE

Investing in Renewable Chemistry

Chem venture conference
2019, April 23rd-24th

SOFINNOVA
PARTNERS FOR LIFE

45 YEARS OF VENTURE CAPITAL

€2B DEDICATED TO LIFE SCIENCES

PARIS BASED
EUROPE FOCUSED – GLOBAL REACH

FUNDING ENTREPRENEURS OF
GAME-CHANGING COMPANIES

Confidential – 2 -

2,5 B€ RAISED DEDICATED TO START UPS IN LIFE SCIENCES

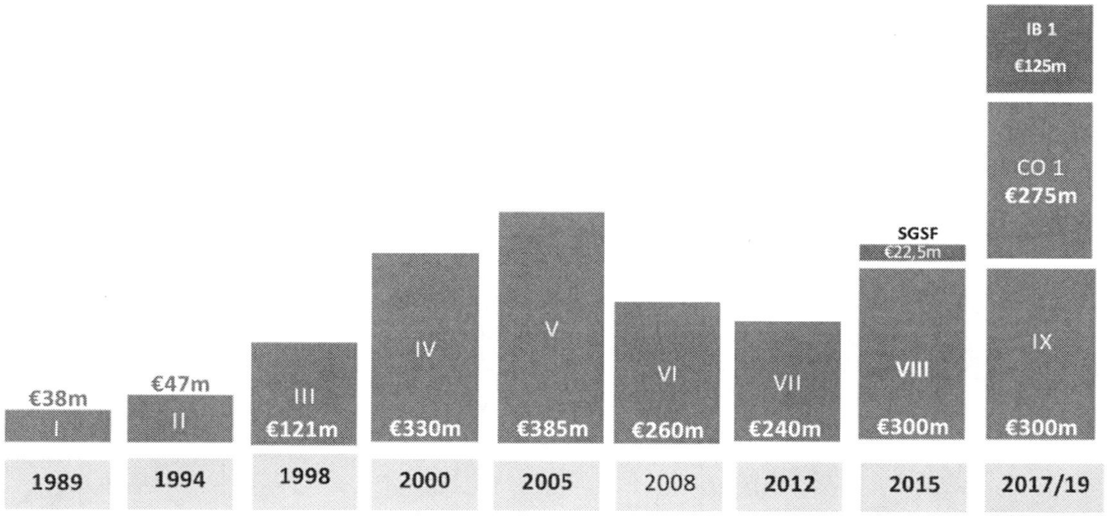

SOFINNOVA PARTNERS

THE SOFINNOVA TEAM

- A team with unique diversity: 13 different nationalities, 12 languages, a rare gender diversity, combining scientific, medical & business skills

SOFINNOVA PARTNERS

SOFINNOVA IB I

Carbon transition : from fossil to renewable

SOFINNOVA IB I

A FUND DEDICATED TO RENEWABLE CHEMISTRY

- A **125 M€** fund
- Investing in **8 to 10 companies**
- Focus on **Seed and Series A**
- In **Europe** and **North America**

Confidential -6-

SOFINNOVA PARTNERS

FOSSIL TO RENEWABLE CARBON in the CHEMICAL INDUSTRY

strong upstream and downstream drivers

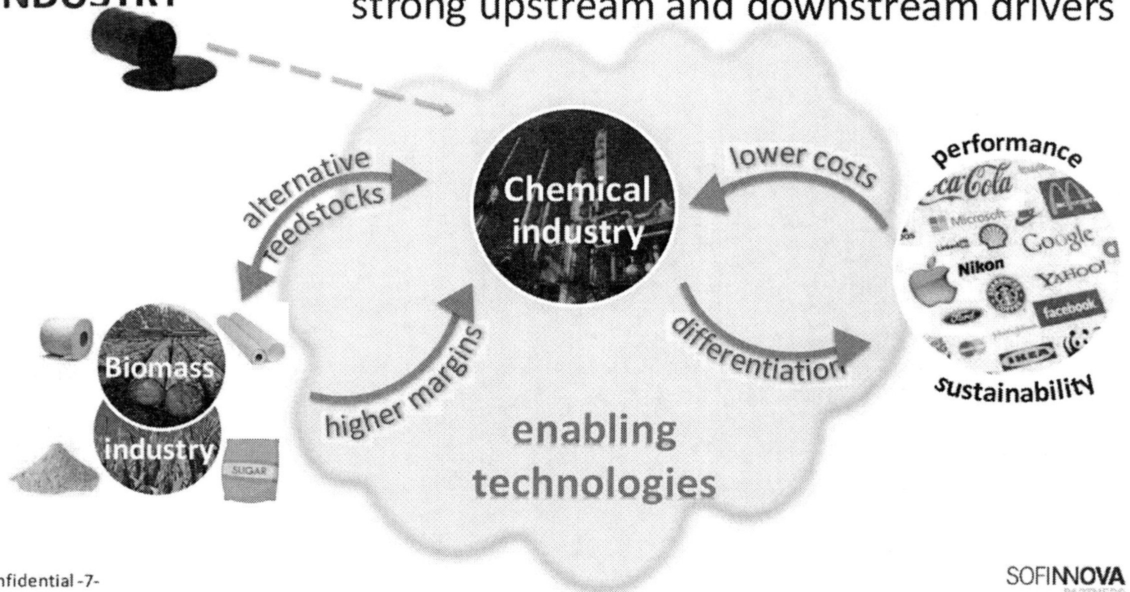

Confidential -7-

SOFINNOVA PARTNERS

INVESTING ALL ALONG THE VALUE CHAIN

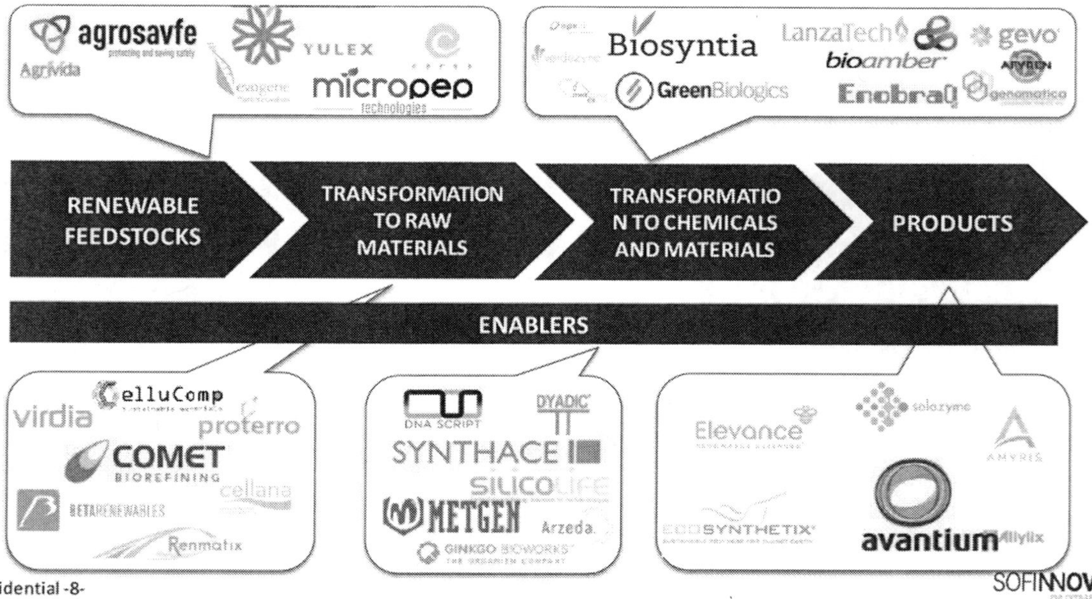

Confidential -8-

SOFINNOVA PARTNERS

ADDRESSING CLIMATE CHANGE

INDUSTRIAL BIOTECH: SAVES 31MT TODAY, POTENTIAL REDUCTION OF 2.5BT CO_2

REDUCTION OF CO_2 EMISSION COMPARED TO FOSSIL EQUIVALENT

- bioamber • **100 % reduction** / fossil succinic acid *
- Green Biologics • **44 % reduction** / fossil origin butanol *
- avantium • **45 to 55 % reduction** / fossil origin PET**
- agrosavfe • **35 % reduction** / fossil origin fungicide***

CO_2 AS FEEDSTOCK FOR CHEMICALS

- Enobraq • yeast using **CO2 as carbon feedstock**

AGRICULTURE WASTE AS FEEDSTOCKS

COMET BIOREFINING — CelluComp

*: Company data
**: Energy and Environmental Science, January 2012
***: "Carbone 4" analysis

Confidential -9-

SOFINNOVA PARTNERS

UNIQUE INVESTMENT EXPERIENCE

agrosavfe, COMET, micropep, Biosyntia, ARVEEN, bioamber, avantium, Green Biologics, METGEN, SYNTHACE, CelluComp, Enobraq, DNA SCRIPT

ACTIVITY	STAGE
SOFINNOVA IB I PORTFOLIO	
Bio-pesticides based on anti-bodies	Early proof of concept
Dextrose and specialty sugars from agricultural waste	Building commercial plant and activity
Micro-peptides for plant regulation and weed control	Early proof of concept
Synthetic biology platform for vitamin production	Product pipeline with proof of concept
Organic acids through consortia fermentation	Building commercial plant and activity
EARLIER SECTOR PORTFOLIO	
Bio-sourced succinic acid and BDO	NYSE IPO in May 2013 Commercial stage at 30 000 T/Y
PEF as bio-sourced alternative to PET	Euronext IPO March 2017 Commercial plant in JV with BASF
Bio-sourced *n*-Butanol	Retrofitting its first commercial plant
Tailored enzymes for biomass degradation	Multiple products developed Early commercialization
Synthetic biology optimization platform	Developing platform through collaborative projects
Nano-cellulose from root vegetable waste	Demo-plant completed, sampling with customers
Renewable chemicals from CO_2	Proof of concept for different chemicals
Synthesis of DNA	Early proof of concept

Confidential -10-

INNOVA PARTNERS

CONTACT

SOFINNOVA PARTNERS

INDEPENDENCE
EQUITY

INDEPENDENCE
EQUITY

Investment Criteria

(Sector) Technologies that promote "Resource Efficiency"

(Stage) Early, demonstrated technology, initial customers

(Geography) U.S.

INDEPENDENCE EQUITY

Resource Efficiency

Cleantech | Agriculture/Ag tech | Materials | Industrial Innovation /Automation/BPO

INDEPENDENCE EQUITY

Portfolio

- Intellihot – energy-efficient water and air heating technology
- NuCurrent – wireless power technology
- 10x Technology – additive manufacturing process technology
- Tagnetics – automated retail technology
- Advanced Diamond Technology – synthetic diamond technology
- Nanostatics – electrospinning of nanofibers
- Swapbox – shipping and delivery automation
- Eatstreet – ordering and delivery technology
- Autonomic Materials – self-healing polymers for coatings and adhesives
- Algal Scientific – use of algae to produce nutrients for animal feed

ALLIED
MICROBIOTA
Biotechnology for a cleaner, safer environment
Funds raised: $600,000 (Grant funding from NSF SBIR Phase I, NYSERDA)
Incubators: CEBIP, PowerBridgeNY, Mcorps
Competitions: Fuzehub finalist, CVC Winners, 76 West Finalist
Conferences: Battelle Conference, Remtec conference, Emerging contaminants summit
Associated research centers: Columbia University, Lawrence Livermore ABPDU, New York Structural Biology Cent
1

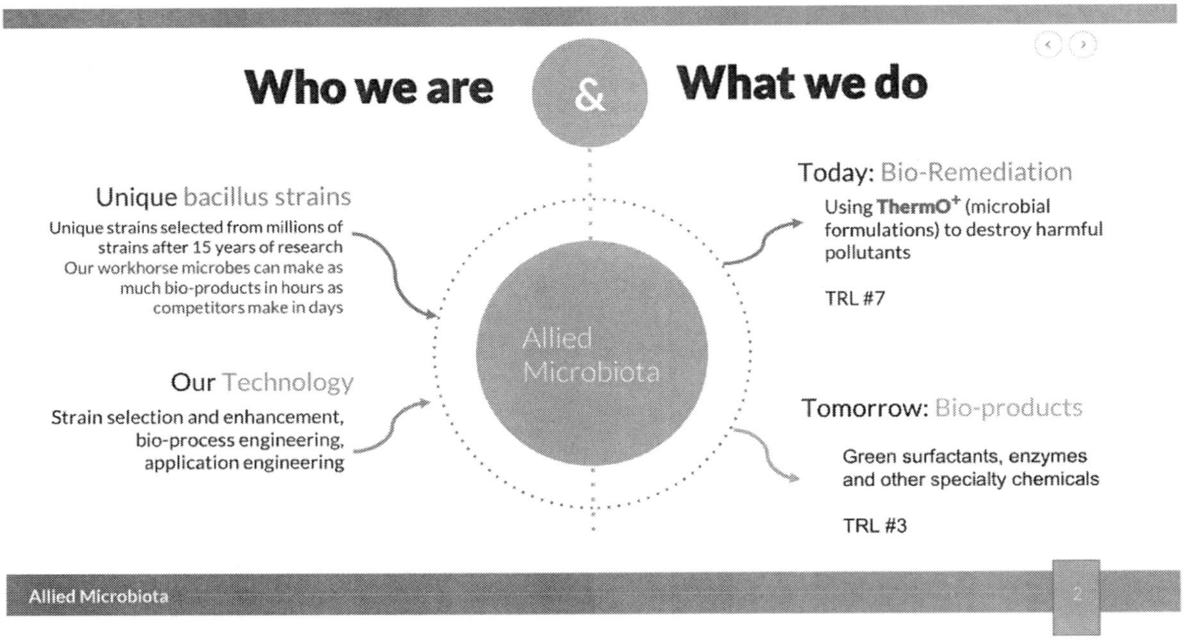
Who we are
&
What we do
Unique bacillus strains
Unique strains selected from millions of strains after 15 years of research
Our workhorse microbes can make as much bio-products in hours as competitors make in days
Our Technology
Strain selection and enhancement, bio-process engineering, application engineering
Allied Microbiota
Today: Bio-Remediation
Using ThermO+ (microbial formulations) to destroy harmful pollutants
TRL #7
Tomorrow: Bio-products
Green surfactants, enzymes and other specialty chemicals
TRL #3
Allied Microbiota
2

35M acres of US land contaminated with toxic organic pollutants

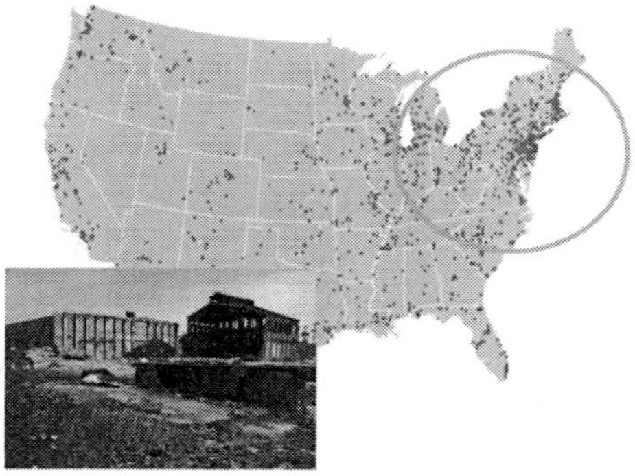

Remediation mandated by federal, state and local governments + 10x increase in real estate value on removing pollutants

Region	Annual Excavated tons (2017)	Market potential for AMB
NE	12 M	$1 B (SAM)
US	100 M	$8 B (TAM)

Global bioremediation market $66B by 2025

Lifetime opportunity of ~$10 Trillion

References: Clean earth marketing department; Market potential assumes pricing of $80/ton for our product

Bioremediation with ThermO+

Leveraging selected bacteria, supported by the right reagents

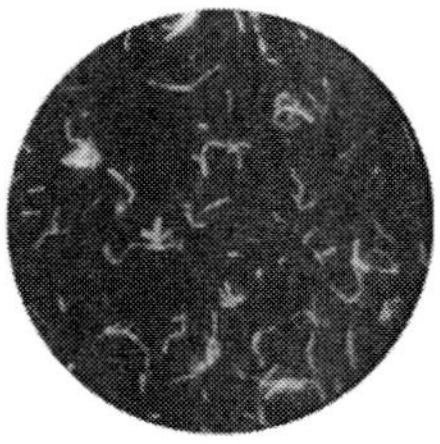

Thermal+: A special blend of naturally occurring, non-pathogenic microbes, containing enzymes, nutrients and accelerants

Implementation: Mix the solution with the soil and incubate at 150°F (65°C) for 2-3 weeks for ex-situ treatment or inject the bacteria into heated soil for in-situ treatment

Bioremediation with ThermO+

Over 20% reduction of contaminants per day

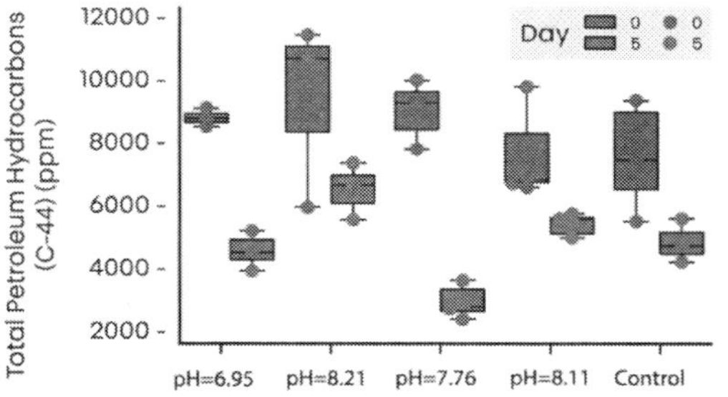

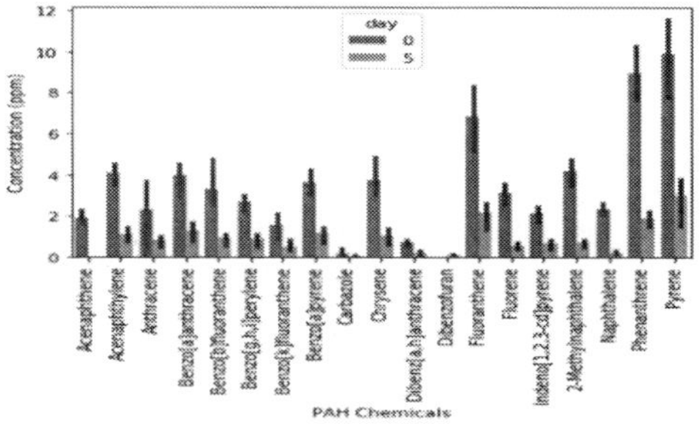

Competitive Landscape

Difficulty to treat

Only a handful incineration facilities and permitted landfills that can treat/store PCB contaminated soil*
AMB ThermO+ family of bio-remediation products can treat all the organic pollutants including PCBs

Business Model

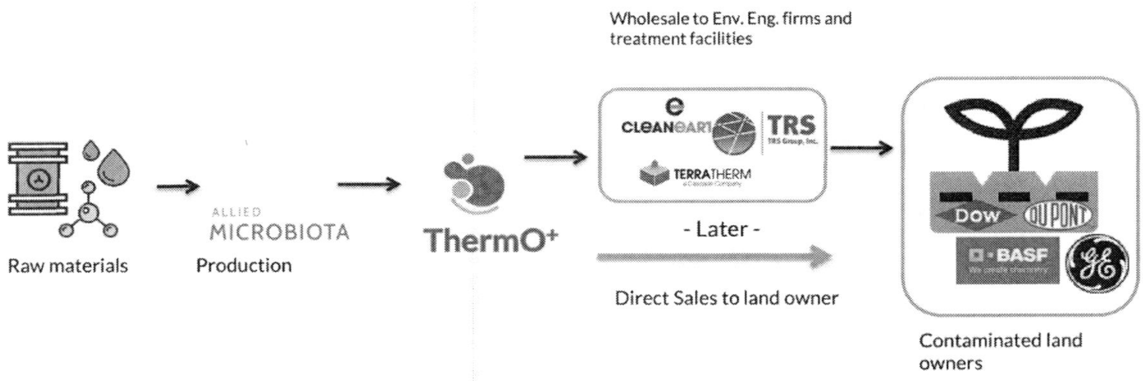

One unit treats one ton of soil*

Allied Microbiota 7

Development Timeline

Proven in the lab, active pilots running successfully as a prelude to sales

2000-2012	2018	2019	2019	2020
Strain development and lab tests	1 ton tech prototype	10 tons field test	1000 ton design	Commercial Operation
	Validated Pac Bac technology on TPH/PAH in field	Determine costs, optimize heat / air / dosing methods at scale	Optimized for capital & operational costs from 10-ton	
Production in shake flasks and 30L	400 Liter Validation	Production Development	Establish facility	
	Demonstrated production technology feasibility	Pilot trial runs at Cornell AgriTech center	Space, Production line, Supply chain	

Allied Microbiota 8

Funding Opportunity: $750k for further testing and scaling up

Use of Funds (From Jan'19)

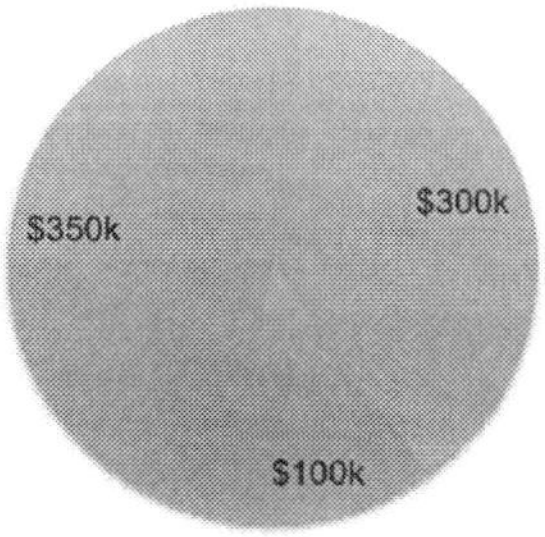

- Medium scale production system setup
- Large scale field design and testing
- Operations and Payroll

Received commitments:
- $130k from NYSERDA received

Pending
- $150k soft commitment from seed VC
- Strategic investment interests from Clean Earth
- $750K from SBIR Phase-II –Applied (June 2019)

<u>**Next round for scale - up**</u>

- $2.5M series A

Team

Ray Sambrotto
CSO
Environmental microbiologist
30+years of experience
At Columbia University

Joel Shertok
Process Advisor
Chemical Engineer
Expert in process scale up

John Santoleri
Advisor
(NYSERDA EIR)
Ex-MD of Warburg Pincus

Frana James
CEO
Industrial engineer, MBA
8+ Years of business development and startup experience

Michael Chin
Dir. of Technology
Application engineer
10+ years of industry experience

Dane Turner
Part - time Chemist
Remediation specialist
20+ years of experience

Katharhy
Associate
MS in environmental engineering

Breakthrough biotechnology
Successful pilot testing
Scaling up technology with a partner

Thank you!

140 58th str., Bldg. A, Ste. 8J,
Brooklyn, 11022

917.680.5436

frana.james@alliedmicrobiota.com

Results from Pilot Tests

- Over 4 tons soil treated in 4 tests at Clean Earth Location in New Castle, Delaware
- 15 treatment conditions tested and 86 samples taken.
- Measured degradation of TPH & 19 specific PAHs, 7 of which are on EPA priority list

Results from Pilot Tests

		CENC2					CENC3					CENC4
Soil mass (lbs):		220	220	220	220	220	220	220	220	220	220	2000
Soil pH:		10.5	10.14	10.12	9.95	10.39	6.95	8.21	7.76	8.11	8.02	5.8
Soil moisture:		16.2	16.2	16.2	16.2	16.2	16.7	16.7	16.7	16.4	16.4	29.5
Other:		control		nitrate		nitrate	non-act.	non-act.	pre-act.	pre-act.	control	
Contaminant	CAS	D1	D2	D3	D4	D5	D1	D2	D3	D4	D5	D1
Acenaphthene	83-32-9						36.3% *	14.6% *	22.5% *	8.3% *	5.5% *	1.6%
Acenaphthylene	208-96-8						25.8% *	18.0% *	25.8% *	16.8% *	4.2% *	11.5% *
Anthracene	120-12-7						27.6% *	10.5% *	20.4% *	7.9%	0.1%	4.7%
Benzo[a]anthracene	56-55-3						22.7% *	10.4% *	19.3% *	7.9%	3.2% *	8.5%
Benzo[b]fluoranthene	205-99-2						25.2% *	10.7% *	16.3% *	8.2% *		
Benzo[g,h,i]perylene	191-24-2						22.1% *	7.9% *	14.6% *	5.8%		4.2%
Benzo[k]fluoranthene	207-08-9						21.5%	7.7% *	21.7% *	8.7%	12.7% *	12.8%
Benzo[a]pyrene	50-32-8						22.5% *	10.7% *	18.2% *	7.0%	2.3% *	0.7%
Carbazole	86-74-8						27.7%					1.2%
Chrysene	218-01-9						25.0% *	11.9% *	21.6% *	9.1%		7.2%
Dibenz[a,h]anthracene	53-70-3						22.7% *	16.3% *	16.1% *	0.6%		0.3%
Dibenzofuran	132-64-9											20.8% *
Fluoranthene	206-44-0						22.8% *	9.3% *	17.6% *	10.8%	0.9%	4.2%
Fluorene	86-73-7						33.1% *	14.6% *	26.9%	17.0% *	5.5% *	4.0%
Indeno[1,2,3-cd]pyrene	193-39-5						22.3% *	8.3% *	15.1% *	5.3%	0.0%	4.5%
2-Methylnaphthalene	91-57-6						35.5% *	9.5% *	19.1% *	3.1% *	0.0%	26.2% *
Naphthalene	91-20-3						44.5% *	30.9% *	34.2% *	23.8% *	19.6% *	21.6% *
Phenanthrene	85-01-8						31.1% *	9.0% *	27.2% *	13.3%	0.6%	9.1%
Pyrene	129-00-0						23.6% *	7.7% *	19.6% *	5.0%	0.0%	6.5%
TPH (C-44)		-8.5%	5.1%	9.7% *	8.0%	-5.8%	13.8% *	7.2% *	21.9% *	6.9% *	8.4% *	4.1%
7 Priority PaHs							23.4% *	10.3% *	18.5% *	7.5%	1.7% *	3.5%
2-Methylnapth. & Napth.							38.3% *	15.4% *	23.5% *	8.6% *	5.7% *	26.5% *

13

Various application approaches for our technology

Treatment method	Capital Expense	Operational Expense	Ease of External Heating	Relative Degradation rate
Land Farming	low-medium	medium	low-high	low-high
Windrow Composting	low	high	low	medium
Passive Aerated-Pile	low	low	low	low
Active Aerated-Pile	low	medium	high	medium
In-Vessel	medium-high	medium	high	high

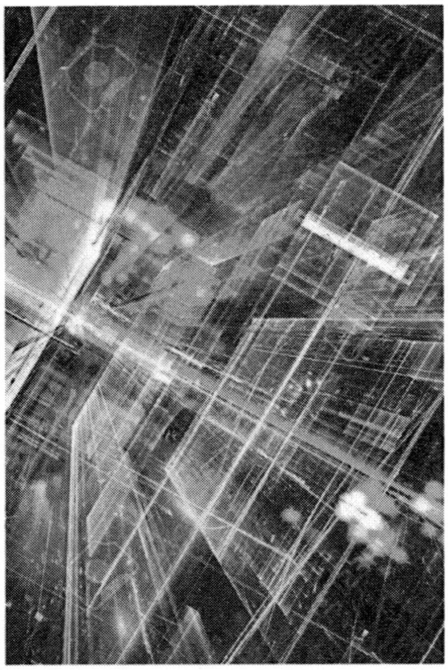

W7energy

ENABLING AFFORDABLE H_2 SOCIETY

CHEMICAL VENTURES CONFERENCE 2019

~~APRIL 23-2~~4, 2019

Prof. Yushan Yan (CEO)
Dr. Santiago Rojas-Carbonell (COO)
Dr. Lan Wang (CTO)

W7energy
Enabling Hydrogen Society

BUSINESS The world's most advanced hydroxide exchange membranes and ionomers

PiperION: a new class of hydroxide exchange polymers that enable the affordable hydrogen society

Scaled up: demonstrated scaling up, required for agile deployment of this new technology

Patent Protection: W7energy's products are protected by six exclusively licensed patents

TRL 6: significantly superior performances than the competitors demonstrated by various companies, research institutes and national laboratories.

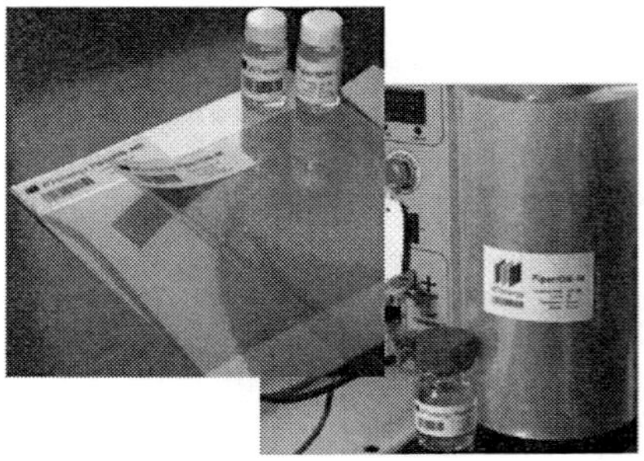

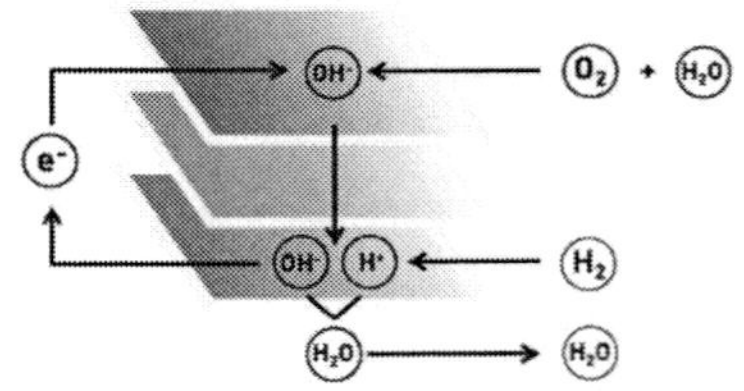

2

MANAGEMENT TEAM

W7energy
Enabling Hydrogen Society

Dr. Lan Wang
CTO
Polymer chemist
Organic and polymer synthesis expert.
Secured $75K in seed funding BlueHen POC

Dr. Santiago Rojas-Carbonell,
COO
MBA
Developer of membrane casting process, management of first customers and business development

Prof. Yushan Yan
CEO
Involved in **several startups** in the capacity of **co-founder**, **Chief scientific advisor**. His patent was licensed to form NanoH2O that was acquired by LG Chemicals

Prof. Shimshon Gottesfeld
Board
Adjunct Professor of Chemical Engineering at the University of Delaware
Leader of fuel cell technology projects
CTO, MTI Fuel Cell
CTO, CellEra

Dr. Mark Mathias
Board
Former Director of Fuel Cell R&D at General Motors
Leader of GM's Battery R&D program

Managing Team | Board Member

Confidential and Proprietary

3

PROBLEM & SOLUTIONS

W7energy
Enabling Hydrogen Society

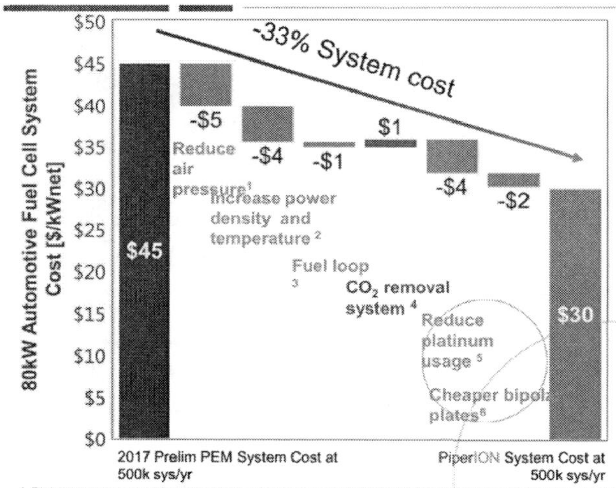

1. Eliminate the expander used for energy recovery, lower compressor cost with lower shaft and motor power, eliminate pre-cooler. 2. Model HEMFC performance with 0.05 Ohm-cm^2 HFR, 400 A/cm^3 ORR (target), 10 μm thick, 1000 A/cm^3 HOR, 3 μm thick (3x current PtRu/C). Use 1.5 bar cathode pressure and 1.4 stoich. 3. Change to "Pulsed ejector with bypass" for fuel loop. This change is projected in the 2025 PEMFC system. 4. ECP module targeting 99.5% CO_2 removal. 5. 30 μg / cm^2 at anode. 1 mg/cm^2 Ag at cathode. 6. Cheaper bipolar plates based on nickel

Unique combination of:

- Mechanical robustness
- Alkaline stability
- High conductivity
- Low production cost

Platform technology: Opens the doors for new materials to be integrated into devices — **<u>Highest collateral benefit on cost reduction</u>**

Confidential and Proprietary

4

TECHNOLOGY & IP

W7energy
Enabling Hydrogen Society

- Advanced polymer design and engineering provide ultimate chemical stability.
- Robust mechanical strength for all applications under extreme conditions.
- Roll to roll membranes and ionomers in alcohol ensures swift introduction in production lines
- Unprecedented high performance on fuel cells and electrolyzers

Invention number	Invention Title
2016-040	Hydroxide Exchange Membranes/Hydroxide Exchange Ionomers Based on Poly (Aryl Piperidinium)
2018-008	Anion and Hydroxide Exchange Membranes and Ionomers Based on Poly-Norbornene-Pyrolidinium
2018-009	Anion and Hydroxide Exchange Membranes and Ionomers Based on Multi-Cation Poly (Aryl Piperidinium)
2018-010	Poly (Aryl Piperidinium) With Stable Cations as Pendants (Combined with UD18-09)
2018-011	Low Tg Thermoplastic Hydroxide Exchange Membranes/Ionomers (Combined with UD18-08)
2019-012	Electrochemical Carbon Dioxide Separator

W7energy's products are protected by six exclusively licensed patents, one of which is already allowed in the USA. The patent protection is currently being nationalized in 11 additional countries.

Confidential and Proprietary 5

MARKET & COMPETITIVE LANDSCAPE

W7energy
Enabling Hydrogen Society

2018 world car market

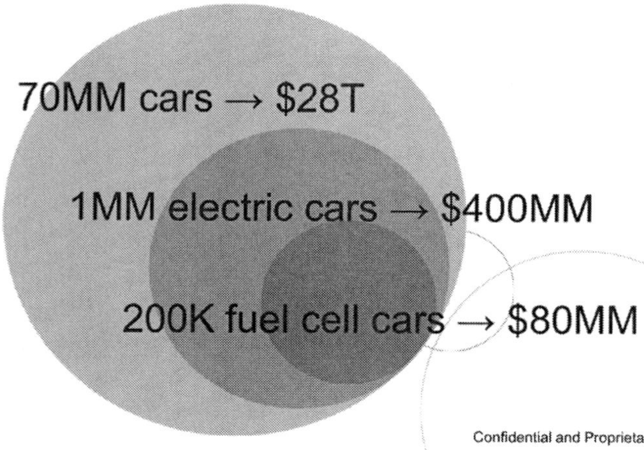

Competitors

fumatech
functional membranes for fuel cells
BWT GROUP

CO_2 Fuel
Dioxide Materials™

HEM: lower performance, brittle membranes

Nafion™ GORE

PEM: excellent performance, corrosive environment, expensive platform

Confidential and Proprietary 6

FINANCING – REVENUE PROJECTION & ASK

W7energy
Enabling Hydrogen Society

Year	2019	2020	2021	2022	2023
Revenue ($)	116	1,581	2,160	10,512	15,684
Gross margin($)	(1,439)	944	1,249	8,588	12,688
EBITDA ($)	(2,106)	268	563	7,600	11,388

Current funding: $4.7MM

- Scale up polymer and film production
- Solidify supply chain and sale channels
- Reach sales of 1000 m^2 of membranes

Ask: $5MM

- Develop PiperION based electrolyzer systems
- Establish electrolyzer production
- Field testing

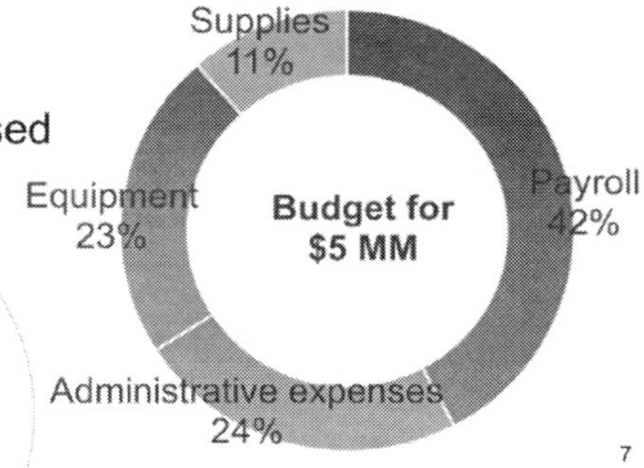

Confidential and Proprietary

7

SUMMARY

W7energy
Enabling Hydrogen Society

Get In Touch

200 Power Mill Rd E500-2404A, Wilmington DE 19803

+ 1 302 200 8619

santiago@w7energy.com

lan@w7energy.com

SAS Nanotechnologies LLC

Innovation of Smart, Advanced and Sustainable Materials for a Better Future

Sumedh P. Surwade
Founder and CEO
e-mail: sumedh@sasnanotechnologies.com
550 S College Avenue, Newark, DE 19716.

SAS Smart Advanced Sustainable Nanotechnologies

Chemical Ventures Conference
April, 23, 2019

COMPANY OVERVIEW

- Innovative materials company
- Founded in 2016
- Located at STAR Campus in Newark, DE
- Technology: Smart microcapsules for self-healing anticorrosive coatings
- TRL 6
- Awards:
  - Akzo Nobel Global Start-up Challenge
  - Swim with the Sharks Competition – New Castle County
  - NSF SBIR Phase I Award

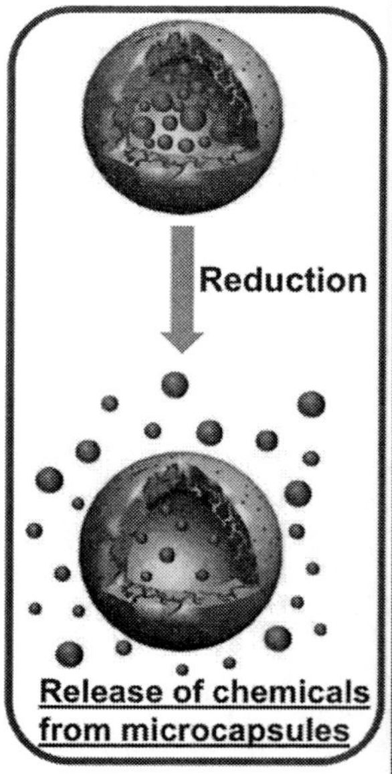

SAS Smart Advanced Sustainable Nanotechnologies

2

SAS NANOTECHNOLOGIES – TEAM

Core Team Members

Dr. Sumedh Surwade
Founder and CEO

Ph.D. in Chemistry
>15 years research experience in nanotechnology

Dr. Kalpana Madgula

Ph.D. in Chemistry
>20 years research experience in materials/catalysis

Advisory Board

Ray Miller
President,
Verdicute Consulting

Dr. Bryan Tracy
CEO,
White Dog Labs

Eric Degenfelder
CEO
US Coatings

Dr. Rakesh Mehta
Counsel,
BakerHostler Law

SAS Smart Advanced Sustainable Nanotechnologies

3

PROBLEM

- Corrosion
  - **One of the biggest problem in the modern world**
    - Economic losses, environmental damage, waste of resources
    - ~4% of US Gross National Product is spent to repair damages caused by corrosion

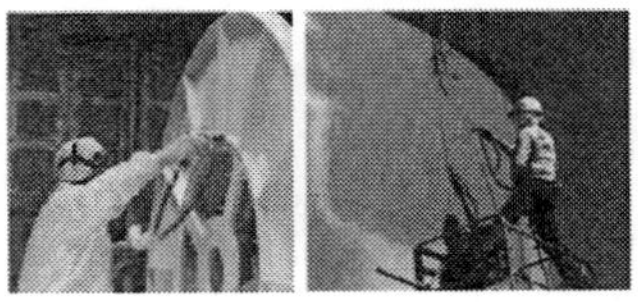

- Anticorrosive Coatings
  - **Problems with Anticorrosive coatings**
    - Most of the coatings contain toxic metal ions such as chromates, zinc, lead etc. – health and environment concern
    - Solvent based – VOC, environmental concerns
    - Coating not effective if there is a scratch on the surface

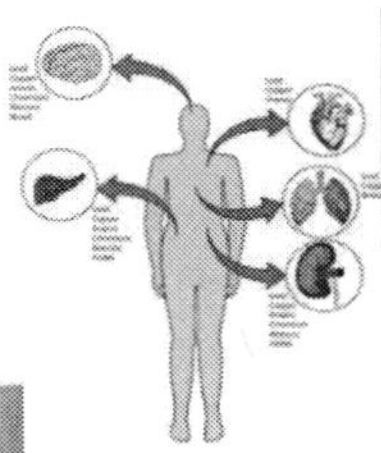

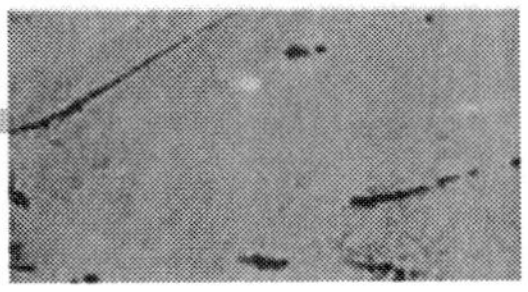

SAS Smart Advanced Sustainable Nanotechnologies

4

SOLUTION: SELF-HEALING ANTICORROSIVE COATINGS

Add smart microcapsule to the coatings

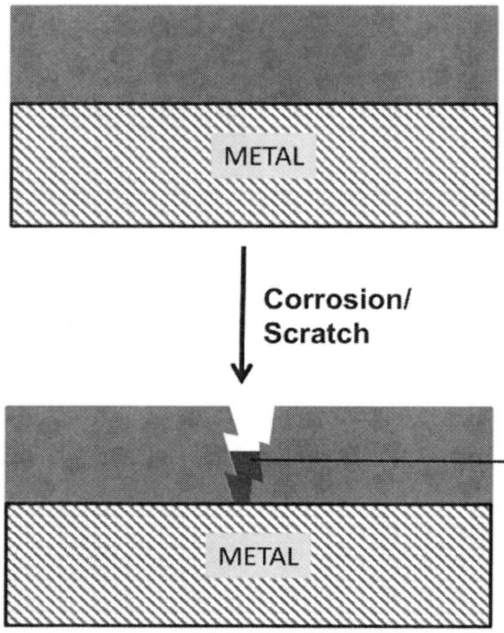

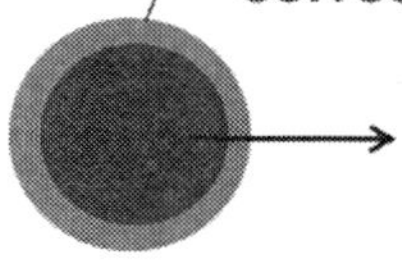

➢ **Value Proposition**

- Heavy metal free (no chromium, lead etc)
- Water based
- Self-healing: coating will heal itself in presence of scratch

Self-healing

In case of scratch, microcapsules gets activated and releases corrosion inhibitor at the corroding site preventing further corrosion

† U.S. Prov. Pat. App. No. 62/698,686; * U.S. Pat. App. No. 16/009,821; PCT App. No. 2018/037796

ANTICORROSIVE COATINGS MARKET

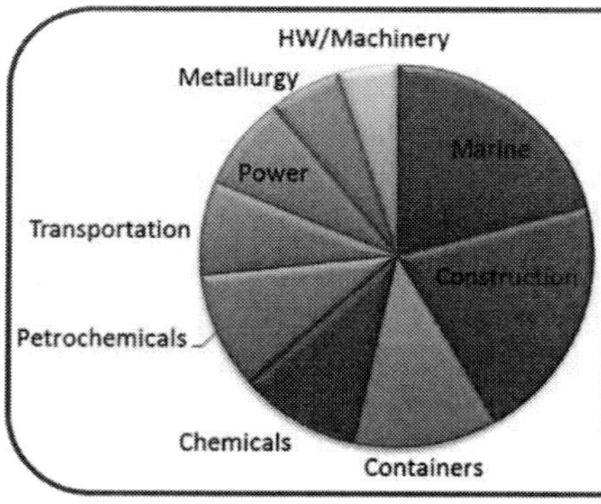

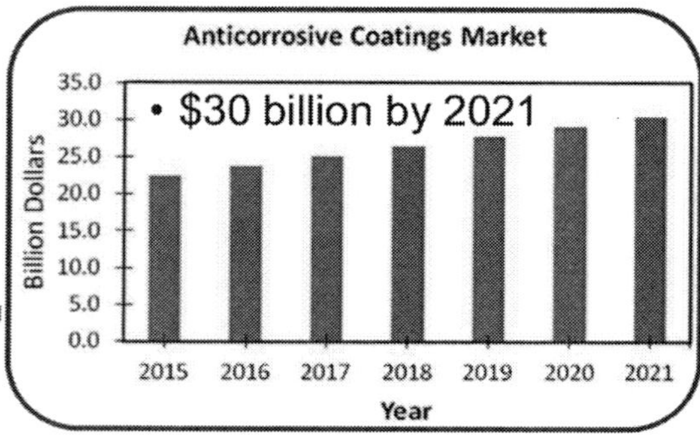

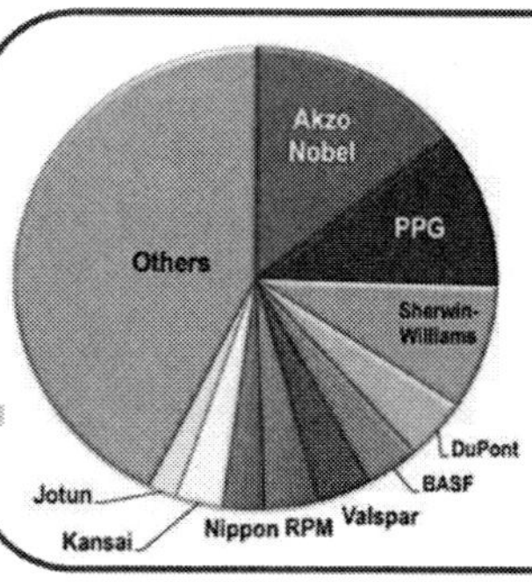

Few companies that own more than 60% of market share

INVESTMENT & FUNDING

Source	Funds	Status	Use
NSF SBIR Phase I award	$225,000	Awarded Jan 2018 – Dec 2018	Salaries, product development, overhead
Angel Investment	$75,000	Obtained in exchange for 15% equity in the company	Equipment, patent attorney fees, salary
SWTS competition Award	$10,000	Awarded	Equipment purchase, overhead
DE SBIR Bridge Grant	$50,000	May 2019	Bridge funding gap until Phase II grant
NSF SBIR Phase II award	$750,000	Seeking (Aug 2019)	Product development, commercialization
Angel/Venture Investment	$400,000	Seeking	Salaries, product development, strategic partnership

SAS Smart Advanced Sustainable Nanotechnologies

7

CONTACT INFORMATION

SAS Nanotechnologies LLC

550 S College Avenue

Newark, DE 19716.

Sumedh P. Surwade, Ph.D.

Founder and CEO

E-mail: sumedh@sasnanotechnologies.com

Tel: 214-235-1008

SAS Smart Advanced Sustainable Nanotechnologies

8

NIS

$IFWS_2$ based high performance additives for: Metalworking, greases, lubricants and other applications

April 23rd 2019

AIChE
The Global Home of Chemical Engineers

NIS

NIS/NML Structure

23 staff members between 2 facilities

NIS USA
Headquarter

NML Israel
100% owned by NIS
subsidiary

Production of concentrates: IF WS2 nanoparticles dispersions in oils

Production of additives: based on Group I,II,III,IV and V oils

Production of grease additive

Synthesis of IF WS2 nanoparticles

NIS Technology overview

- Industrial scale production of novel spherical particles of tungsten disulfide (IFWS2)

- Industrial scale production of **antiwear**, **antifriction** and **extreme pressure** additives:
  - Up to 40% water based – for high performance MWF and coatings
  - Up to 50% oil based concentrates
  - Up to 15% paste formulations (bentonite based)

- Pre-scale up production of inorganic nanotubes (INT), nano and "nanologs":
  - For coatings
  - For catalyst applications (desulfurization process
  - Polymer reinforcement
  - Composites reinforcement

NIS NIS's applications

IFWS2 based additives

Production of greases	Production of MWF	Production of Oils

- ✓ Decreases friction = **increases efficiency by up to 15%, decreases operating temperature, increases life of the lubricant**
- ✓ Decreases wear by 20-37% = **reduces cost of ownership, reduces downtimes**
- ✓ Increases EP properties by 2-2.5x = **reduces cost of ownership, reduces downtimes**

NIS Collaborations

- Since 2014 NIS/NML have established strong collaborations and JDAs with the following companies:

TOTAL · EVONIK (POWER TO CREATE) · Quaker (It's what's inside that counts.) · HOUGHTON · PETRONAS · LUKOIL (OIL COMPANY) · Chemours · DUPONT · bp · Castrol

5

NIS History of NIS/NML and IP

IFWS2 particles discovered. **Prof. Reshef Tenne on a short list for Nobel Prize**

Over 100 patents globally covering:
- ✓ methods of production
- ✓ applications

Techno logy Scale up

NIS acquired NML

NIS developed and scaled up IFWS2 concentration production

Opening of a production facility at NIS and NML

Development of 3 line of products:
- Concentrates/additives for greases
- Concentrates/additives for water based fluids (MWF)
- -Concentrates/additives for industrial oils

Marketing Sales strategy development and implementation etc.

Growth and market development

1992 2002 2013 2014 2015 2016 2017 2018 2019

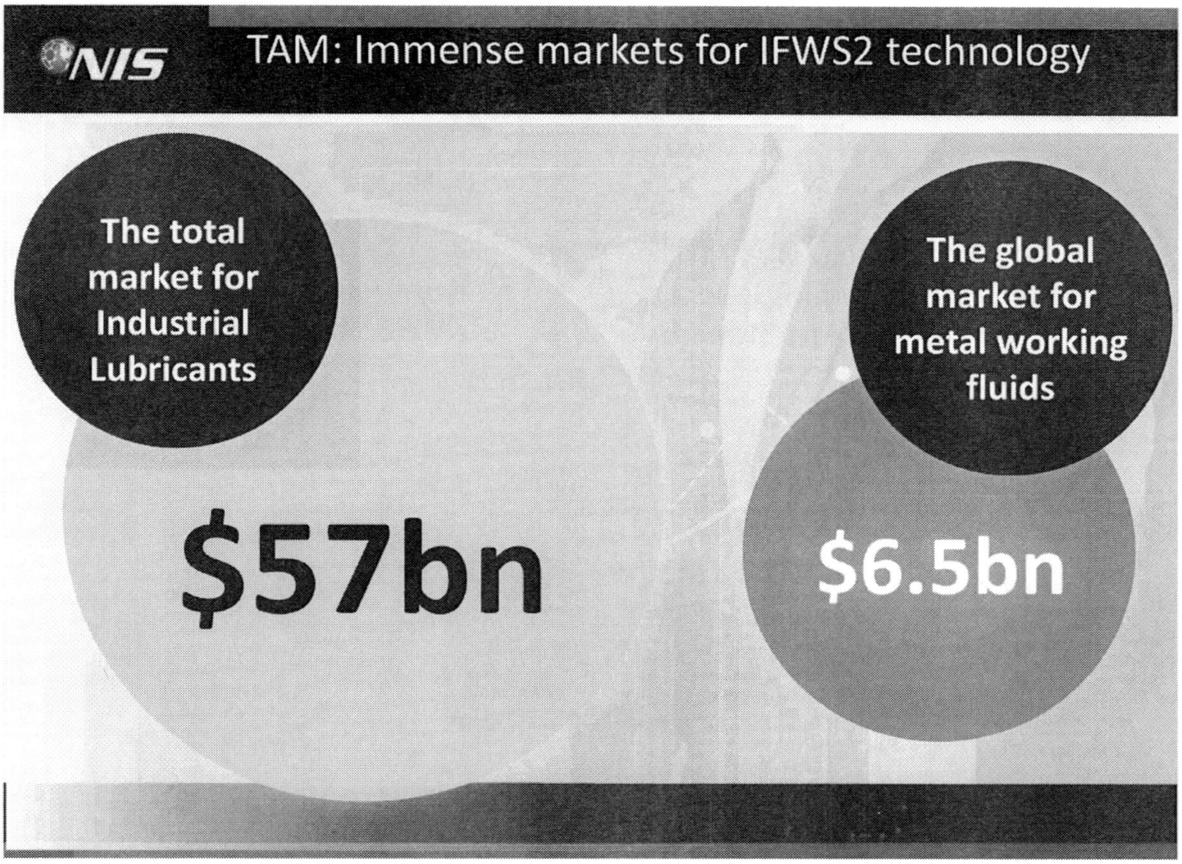
NIS
TAM: Immense markets for IFWS2 technology
The total market for Industrial Lubricants
$57bn
The global market for metal working fluids
$6.5bn

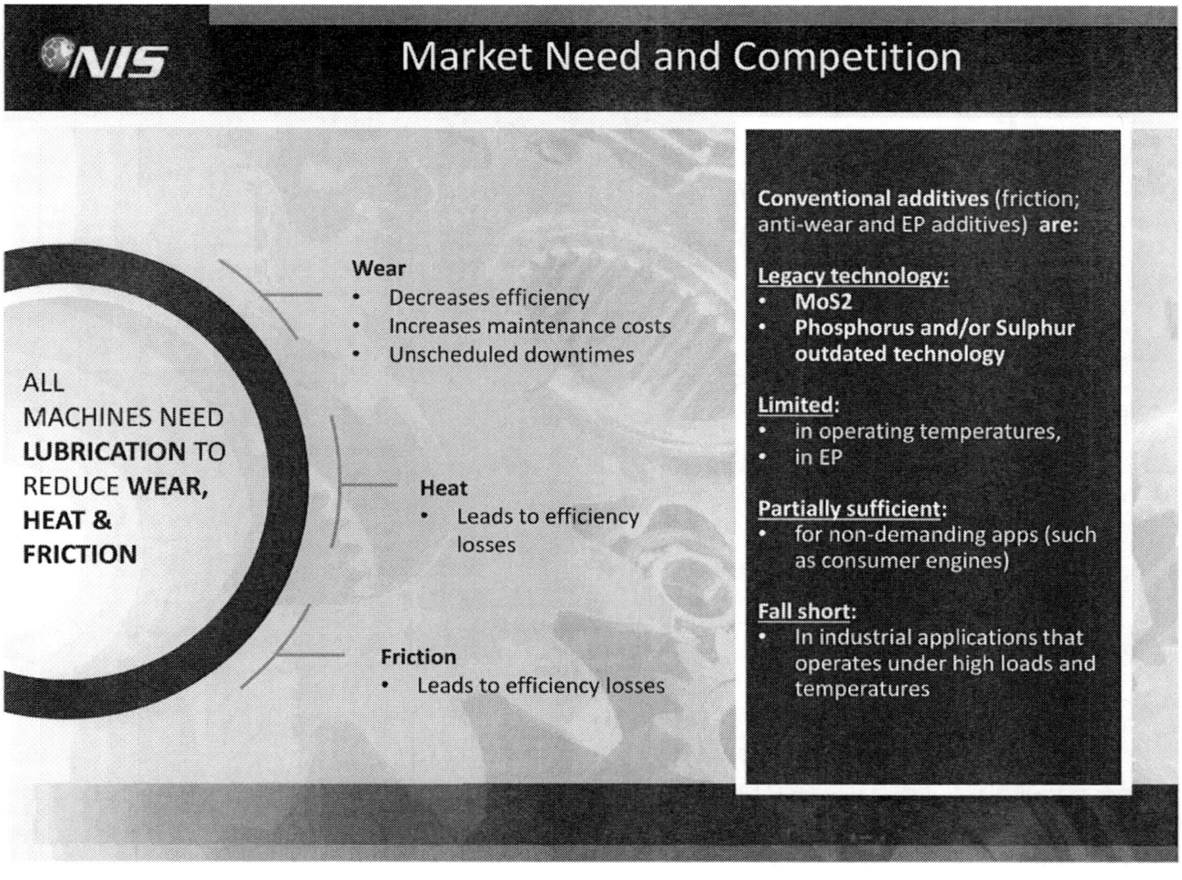
NIS
Market Need and Competition
ALL MACHINES NEED LUBRICATION TO REDUCE WEAR, HEAT & FRICTION
Wear
• Decreases efficiency
• Increases maintenance costs
• Unscheduled downtimes
Heat
• Leads to efficiency losses
Friction
• Leads to efficiency losses
Conventional additives (friction; anti-wear and EP additives) are:
Legacy technology:
• MoS2
• Phosphorus and/or Sulphur outdated technology
Limited:
• in operating temperatures,
• in EP
Partially sufficient:
• for non-demanding apps (such as consumer engines)
Fall short:
• In industrial applications that operates under high loads and temperatures

NIS INT Potential Markets and Applications

There are several big markets have been identified for INTs:

- **Hydrodesulfurization**
- **Composites**
  - **Concrete**
  - **Alloys**
  - **Polymers**

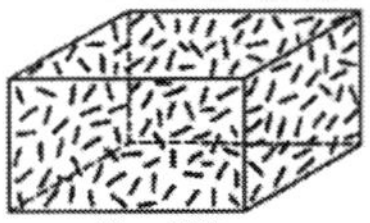

Short Fibers

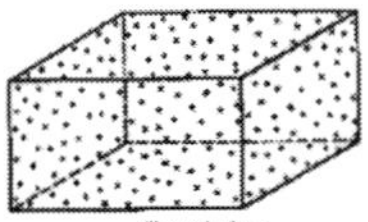

Particles

Currently we are going through identification of potential "champions" for collaboration and tests

NIS Current range of INT Morphologies

Nano-logs:

- 0.3-5 μm wide, up to 30 μm long
- With oxide and/or sulfide core
- >90% pure

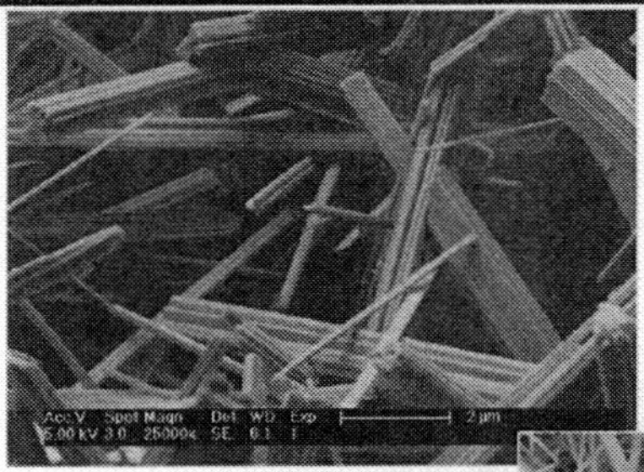

Nano-rods:

- 50-300 nm wide, up to 10μm long
- Hollow or with oxide core
- >90% pure

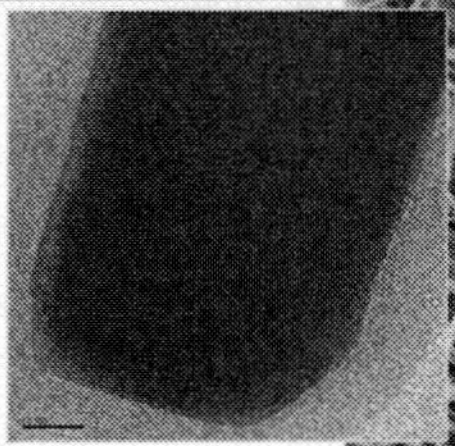

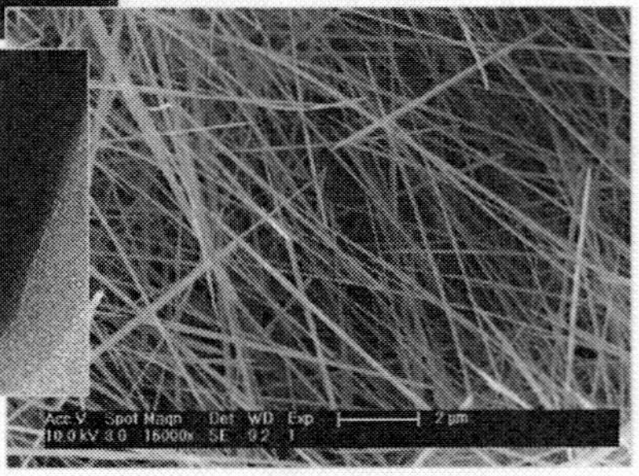

Nano-tubes:

- 10-50 nm wide, up to 10μm long
- Hollow

Currently unavailable as pure phase

NIS

Our know-how

We are the only in the world with expertise of production various shape WS_2 particles and high performance concentrates in commercial scale!

- **Fullerene-like** particles (*IF*-WS_2)– hundreds of **kilo**grams
- **Elongated** nano-particles (INT) – hundreds of grams

NIS

Thank you!

2323 Randolph ave, Avenel, NJ, 07001, USA
1-732-313-0020
www.nisusacorp.com

A Cost Effective, Green Technology for Precious Metals Isolation

2019 Chemical Venture Conference

Roger B Pettman, PhD DSc
roger.pettman@cycladex.com
617 331 1130

Management

Cycladex

Chairman and CEO
Roger Pettman, PhD DSC.
Founder of Cycladex
Co-Founder of Revolymer raising £40m including an AIM IPO
Co-founder Chirex with $147m NASDAQ IPO

Chief Financial Officer
Christine Betts ACA
Former CFO of Microsoft UK with 2000+ employees
Joined Microsoft with 4 UK employees
Responsible for the building of Microsoft Reading Campus.

Independent Non-Executive Director
Lieven De Smedt, eng, MBA, MiF
Business Executive 30 yrs experience
Chairman of the Board at AgroSavfe (AgTech, Belgium)
Chairman of the Board at BioMillenia (BioTech, Paris)

Independent Non-Executive Director
Professor Adrian Boyce
Leader in Applied Geoscience research
4 decades of Scientific Research into Mineral Deposits
Published over 300 peer-reviewed papers
Fellow of the Society of Economic Geologists (SEG) and Geology Applied to Mineral Deposits (SGA)

Executive VP, Technology Transfer
Dale Arnold CCHEM
DuPont Chemical Engineer with 35 yrs experience in plant design, build and commissioning
Engineered titanium dioxide plant in Taiwan from inception to production
Developed and scaled up Cycladex process development

Chief Scientific Officer
Sir Fraser Stoddart, Nobel Laureate
Director for the Chemistry of Integrated Systems at Northwestern University
Knighted by the Queen for his work in nanotechnology
Awards include; the King Faisal International Prize, the Albert Einstein World Award, the Feynman Prize in Nanotechnology and the Tetrahedron Prize for Creativity in Organic Chemistry. Over 1000 publications in the most prestigious scientific journals
Burning ambition to make an impact on society

Independent Adviser
Richard Brown
Chief Commercial Officer at Gabriel Resources Gold Mine (Rosia Montana & Silver Project)
Prior Head of Corporate Finance and COO at Ambrian (Natural Resources Investment Bank)
Funding, regulatory and compliance focus in mining sector
Significant experience in the gold mining market

The Technology vs Cyanide

Cycladex

Cyanide Merrill Crowe Process – 36h Batch Time

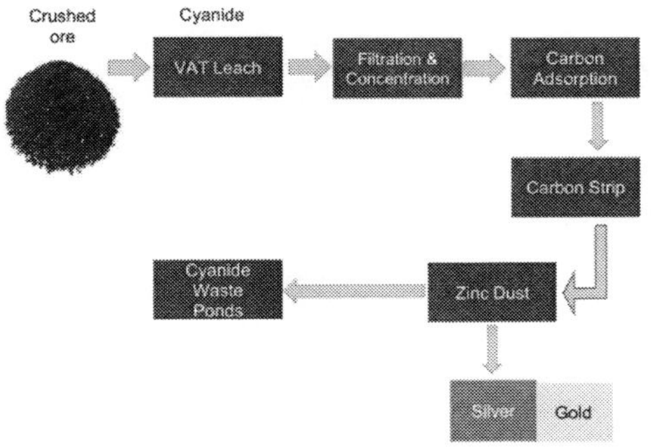

Cycladex Process – 4h Batch Time

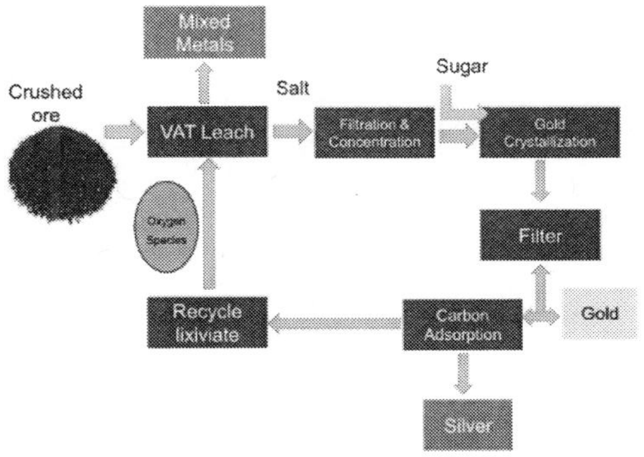

Competition Market and Economics

Cycladex

Competitors

- Dundee Technologies closed loop chlorination
- Enviroleach no data
- Focusing on opportunities where cyanide doesn't work.

Markets

- Tailings with high Fe
  - 8 million tonnes crushed at 2.75gm Au, 18gm Ag
  - Generates 2m tonnes per annum of cinders
- Tailings in Brazil with high Fe content
- Artisinal Miners in Arizona, commercial plant available for scale-up
- Plant in South Africa with refractory ore

Business Model for 1million tpa at 2.75gm of gold tailings

	Cost $
CAPEX Process Plant	50m
OPEX per annum	20m
Gold Revenue	103m
EBITDA per annum	83m

Patents

Cycladex

Patent	Countries / Organisations	Status
Compounds and Methods to Isolate Gold	China, Eurasian Patent Organisation*, Armenia, Azerbaijan, Belarus, Kyrgyzstan, Kazakhstan, Russian Federation, Tajikistan, USA	Granted / Registered
Compounds and Methods to Isolate Gold	Australia, Canada, European Patent Office, Peru, Philippines, Uzbekistan, South Africa	Pending
Gold Process	Australia, Canada, Chile, China, Eurasian Patent Organisation*, European Patent Office, Peru, USA, South Africa	Application Filed
New Process	3 Patents in preparation	

*includes Armenia, Azerbaijan, Belarus, Kazakhstan, Kyrgyzstan, Russia, Tajikistan and Turkmenistan

Funding

Cycladex

Historical

- Cycladex was incorporated in March 2014 as a Delaware corporation
- Seed funding through National Science Foundation (NSF) Small Business Innovation Research (SBIR)
  - Phase I $150k
  - Phase IB $30k
  - Phase II $700k
- Other funding
  - 2015: Comstock Mining $60k cash $240k PIK
  - Angel investing $120k
  - Management $150k

Funding Round

- Seeking $2m to move to profitability

Use of Proceeds

- $250k for pilot plant
- $175k for patent prosecution
- $175k mechanical engineering
- $200k chemical engineering
- $200k Business Development
- $1m SG&A, sales and marketing

Cycladex

Roger Pettman
roger.pettman@cycladex.com

6

Lectrolyst

An electrochemical company for the chemical industry

Greg Hutchings (greg@lectrolyst.com)
President and Co-founder

Raising **$500,000**
Seeking new connections

A scalable electrocatalytic reactor acting as a new platform to make specialty and commodity chemicals without oil or natural gas

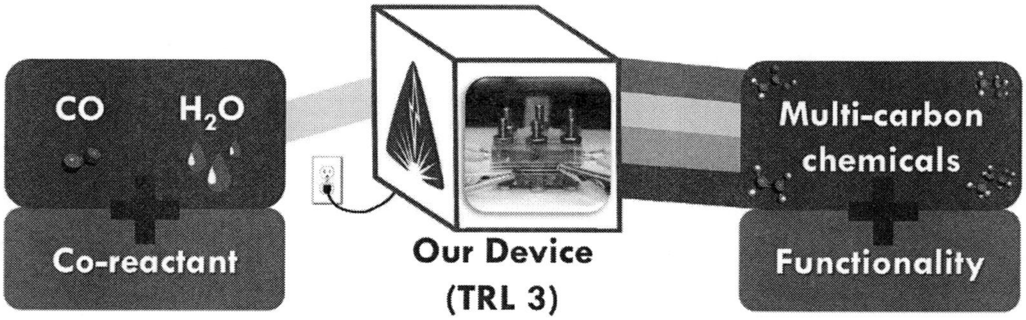

Two core customers

Specialty chemical users

Producers needing CO_2 solutions

Electrocatalysis experts

Dr. Greg Hutchings
Co-founder and President of Lectrolyst
Full-time with Lectrolyst
Postdoctoral Innovation Fellow – U. Delaware
Ph.D. in Chemical Engineering – U. Delaware
19 papers in the field, 1 patent

Prof. Feng Jiao
Associate Professor of Chemical Engineering
World-Class Researcher in Electrocatalysis
Co-founder of Lectrolyst
67 papers in the field, 2 patents

We're building a team to bring this product to market

Lectrolyst 3

10 kg per device per day

tons per day

Core IP from the University of Delaware

✓ CO conversion method

✓ High-rate generation

✓ Specialty chemical recipes

Lectrolyst 4

Meets needs across production scales

Specialty chemical users

Locally generate chemicals on demand (start with sale of C_2-C_4 amides)

Our short-term route:

- Expand specialty library
- Maximize energy efficiency
- Assure long-term stability

Producers needing CO_2 solutions

Transition to CO from captured CO_2 on co-developed plants (device sales)

Instead of the normal pathway:

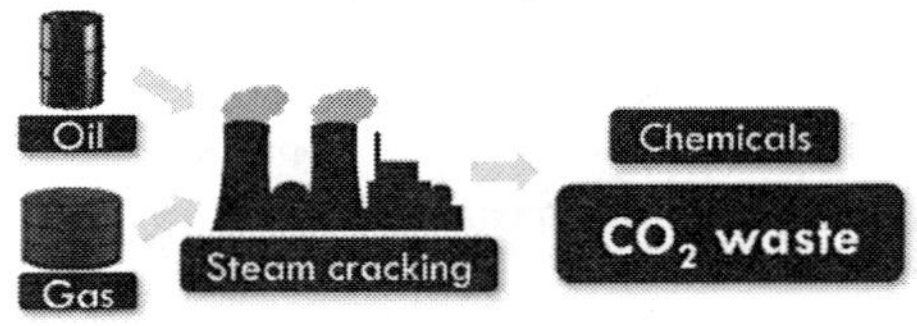

Lectrolyst

Markets and competitive landscape

Initial market: **specialty chemical sales**

Identify compound → Produce on contract → Sell chemicals $100k-1M each

ramp-up ↓ co-development

Long-term market : **co-developed plants**

Addressable reactors market: **$1.8 billion**

		Oil or natural gas	Direct CO_2 E-chem
CCUS compatible	☐	◔	☐
Scalable to local demand	☐	◔	☐
Flexible co-reactants	☐	☐	☐
Short-term profitable	☐	☐	◔

Sources: BCC Research, TechSci Research, Grand View Research, Transparency Market Research

Lectrolyst

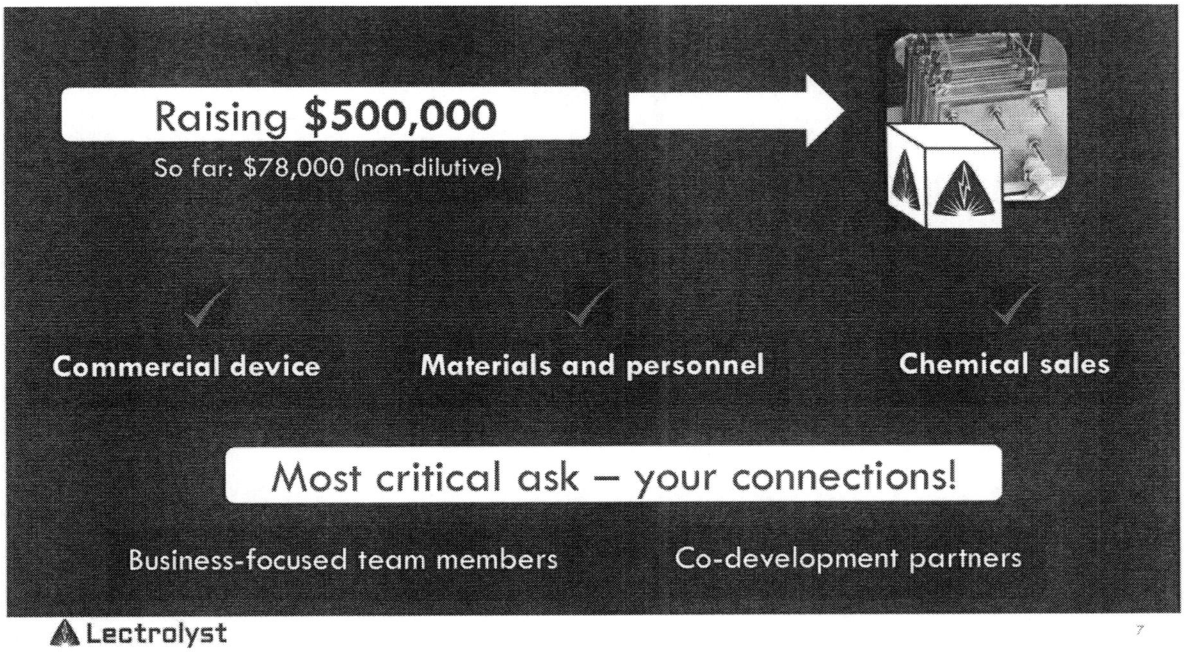
Raising $500,000
So far: $78,000 (non-dilutive)
Commercial device
Materials and personnel
Chemical sales
Most critical ask – your connections!
Business-focused team members
Co-development partners
Lectrolyst
7

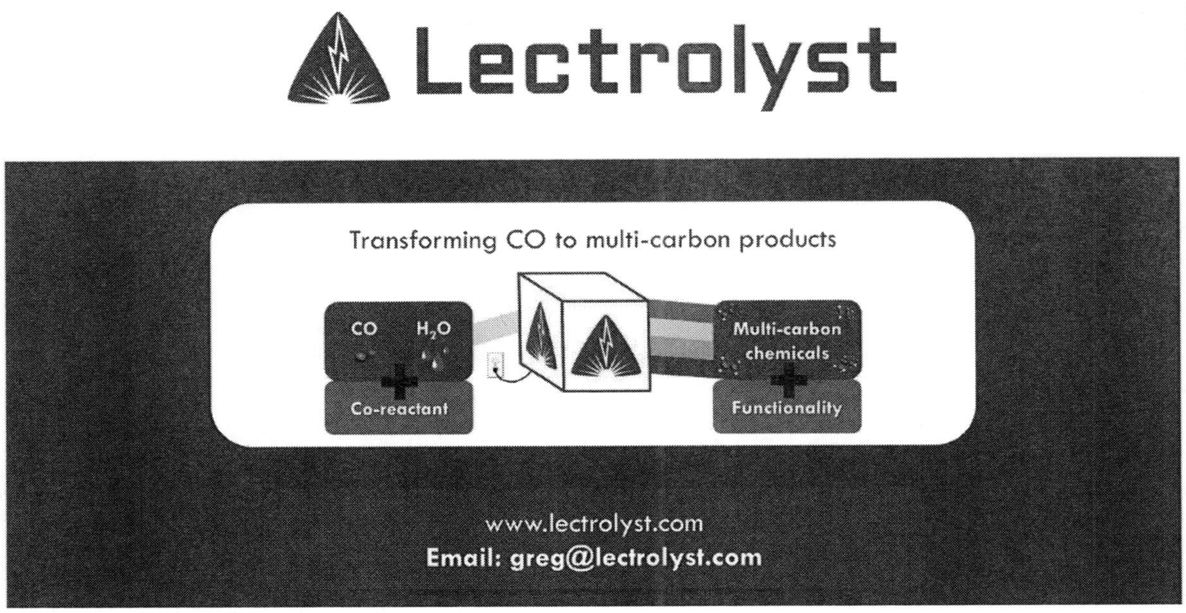
Lectrolyst
Transforming CO to multi-carbon products
CO
H₂O
Co-reactant
Multi-carbon chemicals
Functionality
www.lectrolyst.com
Email: greg@lectrolyst.com

ReliOx Corporation

Eric T. Johnson, CEO
(904)-547-2970
eric@reliox-clo2.com

ReliOx™ **Corporation**
Advanced Chemistry for a Changing World

Affiliation: University of Florida, Office of Technology Licensing

Unlocking Chlorine Dioxide

ClO_2 proven

Decreasing Resistance ↓	TYPE OF MICROORGANISM
	Bacterial Endospores (*Clostridium, Bacillus*)
	Mycobacteria (*Mycobacterium*)
	Non-enveloped, non-lipid viruses (*Parvoviruses*)
	Fungi (*Aspergillus, Stachybotrys*)
	Gram-negative vegetative bacteria (*Escherichia, Pseudomonas*)
	Gram-positive bacteria (*Enterococcus, Staphylococcus*)
	Enveloped, lipid viruses (*Influenza*)

ClO_2
The Pristine Molecule

Decontaminating Agent	Oxidation / Corrosion Potential (V)	
Ozone	2.07	↑ More Corrosive
Peracetic Acid	1.81	
Hydrogen Peroxide	1.78	
Bleach	1.49	
Chlorine Dioxide	0.95	

MOLECULAR SIZE MATTERS

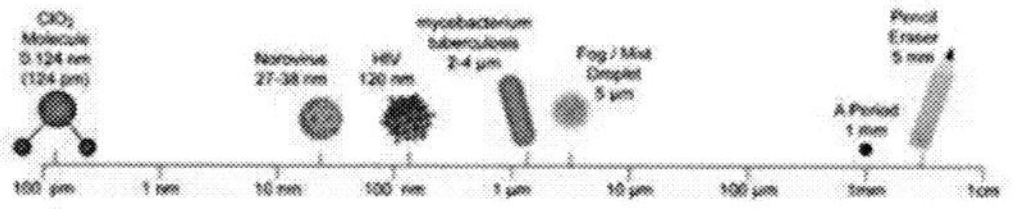

As can be seen in the chart above, the size of a chlorine dioxide gas molecule is 0.124 nm, much smaller than microorganisms and viruses, allowing the gas to easily penetrate into any areas where these microorganisms might be concealed.

ReliOx™ **Corporation**
Advanced Chemistry for a Changing World

Methods of Disinfection

Chemical Disinfectants

- Alcohol
- Chlorine and chlorine compounds
- Formaldehyde
- Glutaraldehyde
- Hydrogen peroxide
- Iodophors
- Ortho-phthalaldehyde (OPA)
- Peracetic acid
- Peracetic acid and hydrogen peroxide
- Phenolics
- Quaternary ammonium compounds

Miscellaneous Inactivating Agents

- Flushing and washer disinfectors
- Other germicides
- Metals as microbicides
- Ultraviolet radiation - UV
- Pasteurization - Heat
- Ozone

Factors Affecting the Efficacy of Disinfection and Sterilization

1) Number and Location of Microorganisms
2) Innate Resistance of Microorganisms
3) Concentration and Potency of Disinfectants
4) Physical and Chemical Factors
5) Organic and Inorganic Matter
6) Duration of Exposure
7) Biofilms

Safe Chlorine Dioxide Availability

1. It has never been done before (NEW)
2. It puts the power of CLO_2 in your hand (POWERFUL)
3. There is no better technology for pathogen control (DISRUPTIVE)
4. Multiple uses across thousands of applications (USEFUL)
5. Safe for human contact (SAFEST)
6. Zero toxic or harmful byproducts (GREENEST)

ReliOx™ Corporation
Advanced Chemistry for a Changing World

"Convenient and Portable Chlorine Dioxide for Your World"

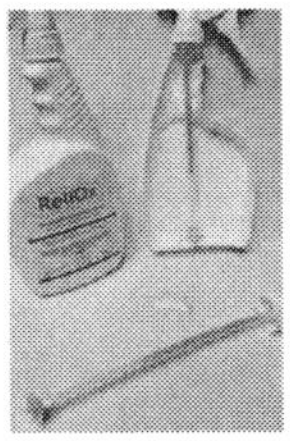

ReliOx™ technology creates the world's best oxidizer

1) In precise aqueous concentrations
2) That can be added/delivered to existing and new devices
3) In spray, pour, wipe, mist, fog, spray, atomize, drip or wash liquids.

ReliOx™ technology eliminates

1) Persistent chemicals and residue
2) Wait time for effectiveness
3) Danger of handling and dispensing harmful chemicals

Available Soon

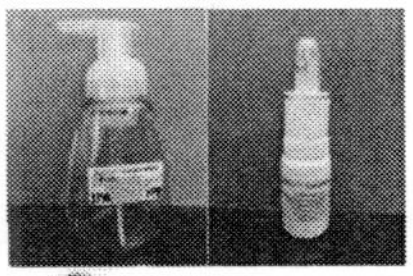

Available Now

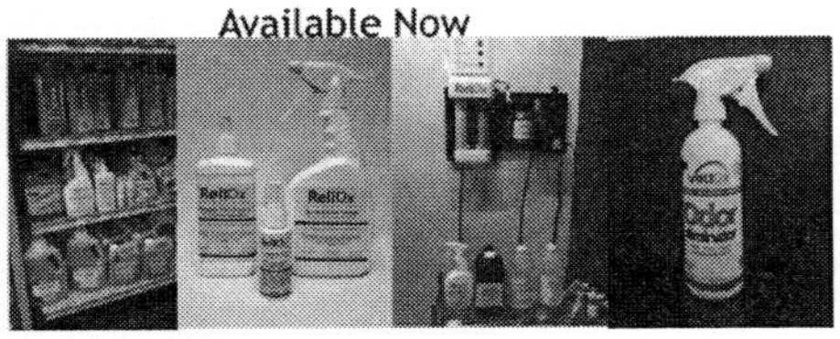

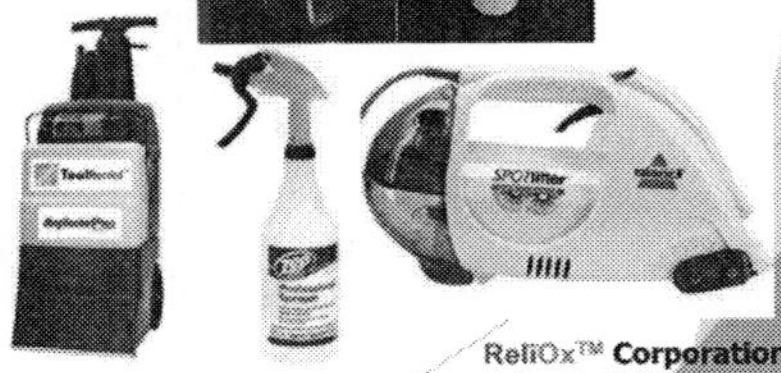

ReliOx™ Corporation
Advanced Chemistry for a Changing World

ReliOx™ Corporation
Advanced Chemistry for a Changing World

100% Patent Protected - Global License

1) U.S. Patent # 7,964,138 - "On-Demand Portable Chlorine Dioxide Generator",5/21/2011

2) U.S. Patent # 8,323,653 - "On-Demand Portable Chlorine Dioxide Generator",12/4/2012

3) U.S. Patent # 9,365,424 - "Chlorine Dioxide Generator using Redox Resin and Adsorbed Chlorite Salt", 6/14/2016

4) U.S. Patent # 9,393,558 - "Preparation of Trihalide Anion Exchange Resins, Reversible Adsorption & Chlorination Reactions Using Ion Exchange Resins, 7/19/2016

5) U.S. Patent # 9,815,051 - "Preparation of Trihalide Anion Exchange Resins" 11/14/2017

6) U.S. Patent allowed # 62/010,534 - Water activated ClO2 Wipes, Sponges

These patents enable the portability and point of use application of chlorine dioxide across thousands segments and millions of uses.

ReliOx™ Corporation
Advanced Chemistry for a Changing World

Management Team

Eric Johnson; CEO
20 year history of sales leadership, distribution and channel development as well as business development working with startups and publicly traded companies such as Mobil, ExxonMobil, Armor Holdings, BAE Systems. Eric is a distinguished military graduate from the United States Military Academy and has degrees in both General Engineering and Electrical Engineering.

Ian Knapp; Founder, President
Ian has held positions such as the Global Director of Marketing for Wheelabrator Water Technologies, CEO of Memtek Corporation and one of three founders of Xydex (GE/Whatman) as well as an Operations Executive for Millipore. Ian earned his degree in Finance from the University of South Florida.

Cristina Knapp; Founder, Vice President, Device Innovation
Former sales and marketing team member at Hoffman LaRoche, Cutter Biomedical Her experience as Marketing Communications Manager for Xydex and Cirqua Customized Water brings insight into product application issues. Cristina received her degree in General Studies and Humanities from the University of Florida.

Bruce Tilley. Ph.D. Vice President, Microbiological Sciences
Manufacturing Manager for Oxthera, Inc. and Manager of Vaccine Production for Aventis Pasteur. His focus on microbiology and the ReliOx product has contributed greatly to the development of the current ReliOx devices and general Chlorine Dioxide capabilities. Bruce is well published and is a graduate of Rutgers University and has his doctorate in Microbiology.

Dawn Tetreault; Director, Channels and Sales Operations
Dawn has sales and business management experience in both large and small technology start-up companies. Successful in achieving results in growing sales while managing multi-million dollar budgets. Strengths in managing sales resources and driving results through channel specific programs.

Michael Thomas, Director, Strategic Relationship Marketing
Mike brings a wealth of experience as an experienced leader and marketer with proven creative and aggressive sales leadership. He was COO of the Blue Sky Group, Inc, HIVIZ Shooting Systems, WellDog as well as an independent business owner at Wyoming Sunrise Ranches, LLC. Mike received his BS in Business, Management, Marketing, and Related Support Services from the University of California, Berkeley, Haas School of Business.

Leaha Wirth, Director, Marketing Communications
Leaha is a disciplined marketing executive who creates winning marketing strategies and programs. Managing partner at BOLT Strategies, LLC and leadership roles at HIVIZ Shooting Systems, LW Marketing, Outdoor Media Network, Armor Accessories and US Olympic Shooting Team Leaha has broad experiences across Many segments. A US Army veteran, she has a BS in Marketing from the University of Phoenix.

ReliOx™ Corporation
Advanced Chemistry for a Changing World

Markets & Competitive Landscape

Enormous Market

	2018	2019	2020	2021	2022
Odor Control	7.5	7.9	8.3	8.7	9.1
Food & Beverage	1.5	1.6	1.7	1.7	1.8
Industrial Water	2.2	2.3	2.4	2.5	2.7
Janitorial	1.5	1.6	1.7	1.7	1.8
Healthcare	1.1	1.2	1.2	1.3	1.3
Total Available Markets	$13.8	$14.5	$15.2	$16.0	$16.8

Market Size in $Billions

Safe & Effective?

SWOT

Strengths (+)	Weaknesses (-)
1. Patents 2. Great team 3. Existing chemical (known) 4. Simple & aggressive plan 5. Simple production 6. Repeat users 7. Safer product 8. Most effective product	1. Strong incumbents 2. Unknown technology 3. No-scent
Opportunities (+)	**Threats (+)**
1. Increased pandemic threats 2. Competitors lack of alternatives 3. Flat surface, air, next to skin, internal uses 4. New technologies in development 5. International markets	1. Focus 2. Consumer acceptance 3. FDA/EPA acceptance 4. Competitor "me too" copycats 5. Product misuse & publicity

Competition	ClO_2	Quats	Bleach	Peroxide	Ozone
No Harmful Residues	✓	X	X	✓	✓
Kills All Pathogen types	✓	X	✓	✓	✓
Not Corrosive	✓	✓	X	X	X
Low Effective Concentration	✓	✓	✓/X	X	X

ReliOx™ Corporation
Advanced Chemistry for a Changing World

Financing - Investment Sought & Use of Funds

ReliOx™ Corporation seeks funding of $2MM

Funds required for:

1) Near term production, marketing and sales
2) FDA product certification
3) Product / business development activities & expenses

Current Investing

Founders	$108,800
Convertible Debt	$517,140
Long term Debt	$53,898
Payables	$86,416
Total	$766,254

Pro Forma Profit and Loss	2019	2020	2021	2022	2023
Sales	$267,000	$3,917,000	$18,083,000	$23,297,000	$31,779,000
Total Cost of Sales	$185,683	$1,671,412	$6,352,758	$8,222,227	$11,051,938
Gross Margin %	30.46%	57.33%	64.87%	64.71%	65.22%
Total Operating Expenses	$1,345,840	$2,836,430	$3,461,625	$3,553,711	$3,740,916
Profit Before Interest and Taxes	($1,264,523)	($590,842)	$8,268,617	$11,521,062	$16,986,146
EBITDA	($1,264,523)	($590,842)	$8,268,617	$11,521,062	$16,986,146
Net Profit	($1,314,294)	($670,613)	$5,711,193	$7,987,904	$11,813,463
Net Profit/Sales	-492.24%	-17.12%	31.58%	34.29%	37.17%

Current Activity

1) Odor Market - Pet Wholesale / Retail
   a) Global Pet Expo launch (Orlando, FL) - March 2019
      a) License, Private Label, Distribution
   b) Odor Market - Commercial Facility
      a) Installed units (Model 48) (48gl / 300ppm)
2) License Opportunities
   a) Church & Dwight (phase 5) Engaged (L/I/A)
   b) Unilever (phase 4) Engaged (L)
   c) Colgate Palmolive (phase 2)

ReliOx™ Corporation
Advanced Chemistry for a Changing World

ReliOx Corporation

Eric T. Johnson, CEO
(904)-547-2970
eric@reliox-clo2.com

ReliOx™ Corporation
Advanced Chemistry for a Changing World

Snapshot of Chlorine Dioxide Production Methods

Traditionally, chlorine dioxide for disinfection applications has been made by one of three methods using sodium chlorite or the sodium chlorite - hypochlorite method:

1. $2\ NaClO_2 + 2\ HCl + NaOCl \rightarrow 2\ ClO_2 + 3\ NaCl + H_2O$

   or the sodium chlorite - hydrochloric acid method:

2. $5\ NaClO_2 + 4\ HCl \rightarrow 5\ NaCl + 4\ ClO_2 + 2\ H_2O$

   or the chlorite - sulfuric acid method:

3. $8ClO_2^- + 4H_2SO_4 \rightarrow 4ClO_2 + 2HClO_3 + 4SO_4^{2-} + 2H_2O + 2HCl$

All three sodium chlorite chemistries can produce chlorine dioxide with high chlorite conversion yield, but unlike the other processes the chlorite-sulfuric acid method produces completely chlorine-free chlorine dioxide, although it suffers from the requirement of 25% more chlorite to produce an equivalent amount of chlorine dioxide. Alternatively, hydrogen peroxide may be efficiently used in small-scale applications.

ReliOx™ Corporation
Advanced Chemistry for a Changing World

Snapshot of Chlorine Dioxide Production Methods (cont.)

Other

Very pure chlorine dioxide can also be produced by electrolysis of a chlorite solution: (Systems have failed to live up to theoretical promise.)

4) $2\ NaClO_2 + 2\ H_2O \rightarrow 2\ ClO_2 + 2\ NaOH + H_2$

High-purity chlorine dioxide gas (7.7% in air or nitrogen) can be produced by the gas-solid method, which reacts dilute chlorine gas with solid sodium chlorite:

5) $2\ NaClO_2 + Cl_2 \rightarrow 2\ ClO_2 + 2\ NaCl$

In the late 1990 until present, the inventors of item 4 produced another ClO_2 method:

6) NaClO2 => IX + > hypochlorous acid => Catalyst = ClO2

This method produced high purity ClO2 but is complicated by sophisticated control systems.

Since 1999 a growing proportion of the chlorine dioxide made globally for water treatment and other small-scale applications has been made using the chlorate, hydrogen peroxide and sulfuric acid method, which can produce a chlorine-free product at high efficiency.

RefiOx™ Corporation
Advanced Chemistry for a Changing World

Small Quantity Generators

Alternative methods have been developed to produce small (one liter to 5 Gallon) quantities of ClO2.

These use two primary methods:

1. **Two Part chemical mixing kits:**
   1) Acid: Acids (sulfuric or hydrochloric or citric or lactic) Dry powder or liquid.
      OR, strong oxidizers such as dichloroisocyanurate, and
   2) Sodium chlorite: Dry powder of liquid

2. **Pouches/ Sachets:**

Several companies utilize **pouches** to contain the powders. This allows water to be added to start the chemical reaction. The use of specialized membranes allows the production of ClO_2 within the pouch and its release over an extended period of time. Specialized membrane materials are sometimes used.

3. **Tablets:**

Composed either of:
1) Sodium chlorite 20.8%; Sodium dichloroisocyanurate dihydrate 7.0% plus an inert tablet filler, or
2) Sodium chlorite and dry acid plus tablet filler.

RefiOx™ Corporation
Advanced Chemistry for a Changing World

Small Quantity Generators (cont.)

a) ALL of the methods listed are attempts to make ClO_2 readily accessible.
b) All methods that use Acids produce low pH, acidic forms of chlorine dioxide.

Note: Frequent attempts to use ClO_2 from these products has resulted in the myth that ClO_2 itself is acidic.

Unlike all the processes used before, the RelīOx technology does not acidify sodium chlorite. Instead the RelīOx process splits the chlorite molecule - rendering it into chlorine dioxide with a remainder of sodium bromide ($NaBr_2$).

The process of creating ReliOx CAP media binds a dense coating of sodium tri-bromide onto a bed of ion exchange resin. The RelīOx process works when a small amount of sodium chlorite, dissolved in water, is passed through a column containing RelīOx CAP media. The NaBr and ClO_2 produced exit the tri-bromide column together. A second step removes the bromide and excess sodium chlorite.

RelīOx is simply a better way to make chlorine dioxide

All RelīOx ClO_2 produced:

- is liquid (H_2O) in form
- is pH neutral
- is not acidic
- contains no acidic products
- requires no chemical mixing
- requires no storing or transporting of acids
- requires no dilution
- uses dilute sodium chlorite solutions (< 5%)
- are made from shelf stable components (RelīOx resin has multiyear stable shelf life)

RelīOx™ **Corporation**
Advanced Chemistry for a Changing World

MesoMat

Stretching Material Possibilities

Paul Fowler and Sukhbir Kalirai

1

Business Summary

Proprietary **stretchable, lightweight, conductive** plastic wires which can be stretched like a rubber band

Platform technology that is **customizable, cost-effective** and **scalable**

- **Stretchable circuitry** for industrial smart materials
- **Sensor to comprehensively monitor** fatigue, failure and fracture in critical material systems

TRL Level: 4-5

2km spool of MesoMat conductive fibers

MesoMat

2

Management Team

Kari Dalnoki-Veress, PhD
Co-Founder

Professor,
McMaster University

100+ publications,
6000+ citations,
industrial consulting

Sukhbir Kalirai
Co-Founder

Director of Business Development at a 3G Capital company

Founder CanLeads

Paul Fowler
Co-Founder

Head of R&D at Validere, YC S16

Extensive international research experience

Clare Armstrong, PhD
VP Product Dev.

Manager of R&D at EKS

NSERC scholar, 15+ publications

3

Challenges Being Addressed

✘

Current sensing solutions and monitoring are:

- **NOT** easy to implement without impacting material integrity
- **NOT** able to monitor entire structure
- **NOT** simple to incorporate within structure
- **NOT** cost-effective

MesoMat

✔

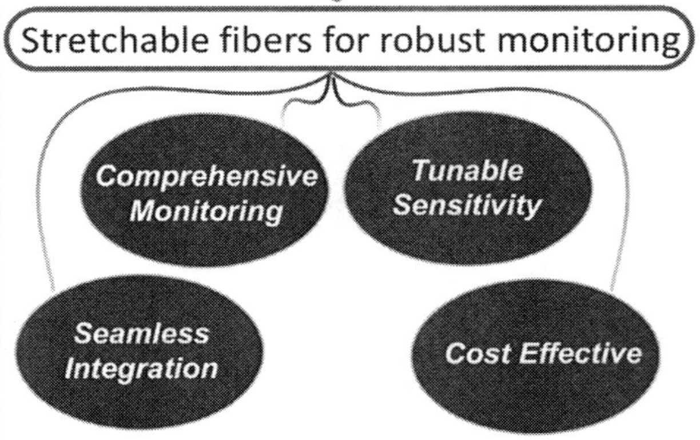

<u>MesoMat</u> provides *both* point sensing *and* global sensing enabling:

Precise identification | *Real-time monitoring* | *Minimal downtime* | *Reduced maintenance*

4

Technology & IP

Multiple patents (PCT pending) covering both core technology, including fiber materials, production processes and technology applications
All IP owned by MesoMat

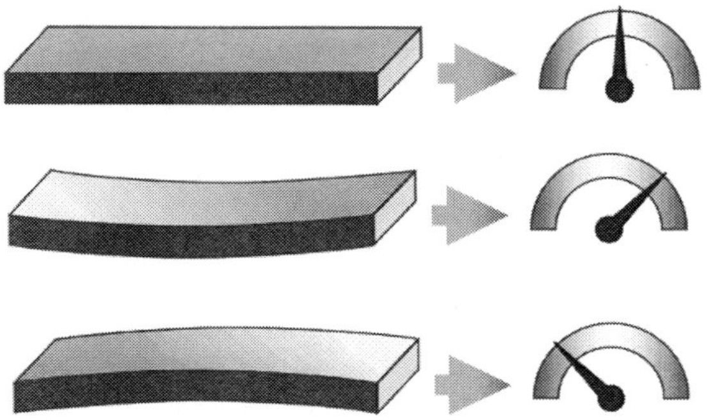

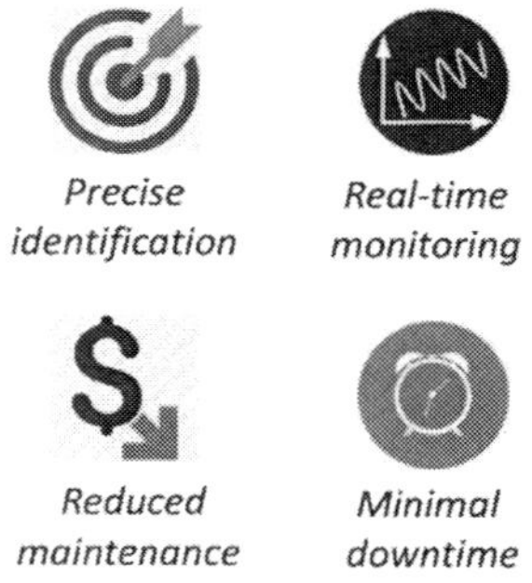

***Sensing fibers** seamlessly embedded into materials to enable intelligent sensing and full SHM solution*

MesoMat

5

Market Landscape

$5B
Intelligent self-sensing composite

- Structural health monitoring (SHM) in critical material systems
- Composite materials
- Wear testing in large assets

Partnership model to build full sensing platform:
Continuous real-time monitoring
with precise failure identification

INTEGRATE MONITOR ANALYZE DISPLAY

Platform	Cost-Effective	Global monitoring	Easy to implement	No structural impact	Accurate
MesoMat	✔	✔	✔	✔	✔
Fiber Bragg grating	✘	✘	✘	✘	✔
Standard strain gauge	✔	✘	✔	✘	✔

6

Financing

Q1 '19

US $750K

Seed Round

✔

Milestones

- *Secure $100k in pilots and convert pilots into recurring revenue*
- *Secure co-development agreement in SHM*
- *Increase production by 500% and lower costs*

Q4 '20

✔

Series A financing to scale business

MesoMat

7

Summary

Proprietary **stretchable, lightweight, conductive** plastic wires which can be stretched like a rubber band

Platform technology that is **customizable**, **cost-effective** and **scalable**

- **Stretchable circuitry** for industrial smart materials
- **Sensor to comprehensively monitor** fatigue, failure and fracture in critical material systems

Manufacturing capabilities, technical expertise and MesoMat's resources provide a pathway to scale solution

Contact Info:

Mesomat.com

info@mesomat.com

MesoMat

8

MesoMat

Mesomat.com

9

Appendix

10

Fiber Sensors: Specs

- **Lightweight:**
  - ~10x lighter than comparable fiber sensor solutions (e.g. optical fiber based sensors)
  - 10 μm to 1 mm in diameter
- **Sensing optimization:**
  - Global & local monitoring: global sensing but also tunable to extreme sensitivity where high local sensitivity is required
  - Piezo-resistive fiber is highly tunable
  - Ultra-thin fibers don't initiate delamination or failure in composites
  - Simple data readout
- **Customizable and tunable in a cost effective way**
- **Durability/reliability:**
  - Fiber sensors function over broad temperature ranges
  - Chemically resistant (e.g. epoxies)

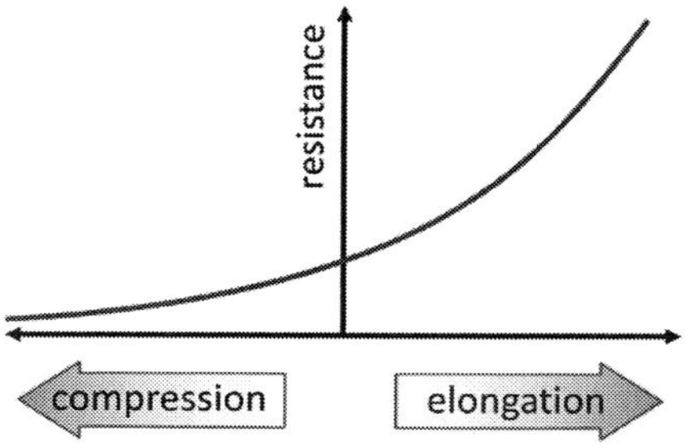

11

Conductive Fibers: Specs

- **Diameter:** 50 μm to 1 mm
- **Resistance**: ~1 Ω/cm, depending on diameter, highly tunable
- **Data transfer capabilities**
- **Resistance vs elongation:**
  - Highly tunable and customizable
  - With ~25% elongation, ~20% increase in resistance, before returning to original value when fiber is relaxed
  - Small increase in resistance does not impede practical use
- **Durability:**
  - Fibers tested over 10k stretch cycles
  - Reliable operation up to 150°C, higher temperatures can be achieved
  - Can be tuned to survive exposure to various solvents/chemicals
  - Laundry tested with no degradation
  - Resistance of fibers remains unchanged (~6 mo) with no degradation

MesoMat

12

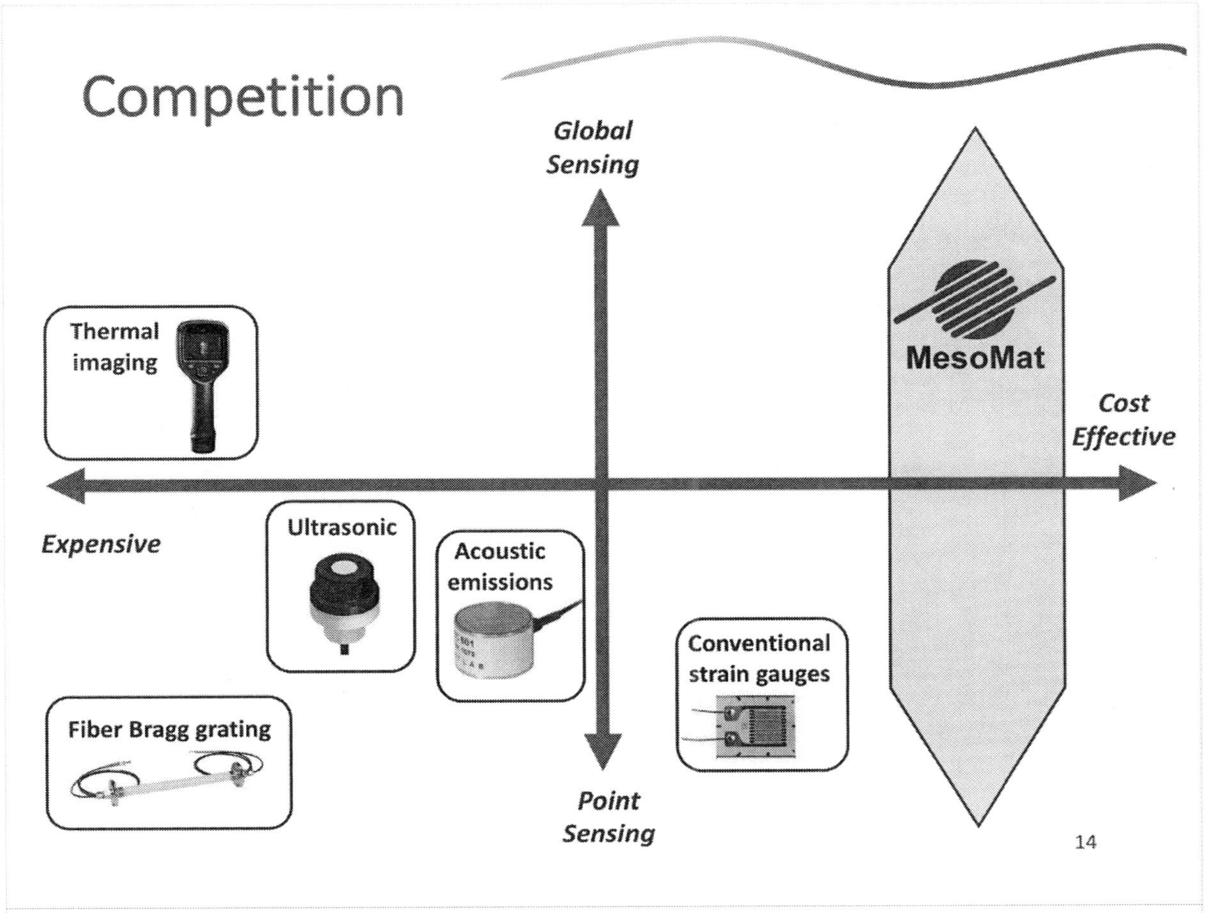
Competition
Global Sensing
Point Sensing
Expensive
Cost Effective
Thermal imaging
Ultrasonic
Acoustic emissions
Conventional strain gauges
Fiber Bragg grating
MesoMat
14

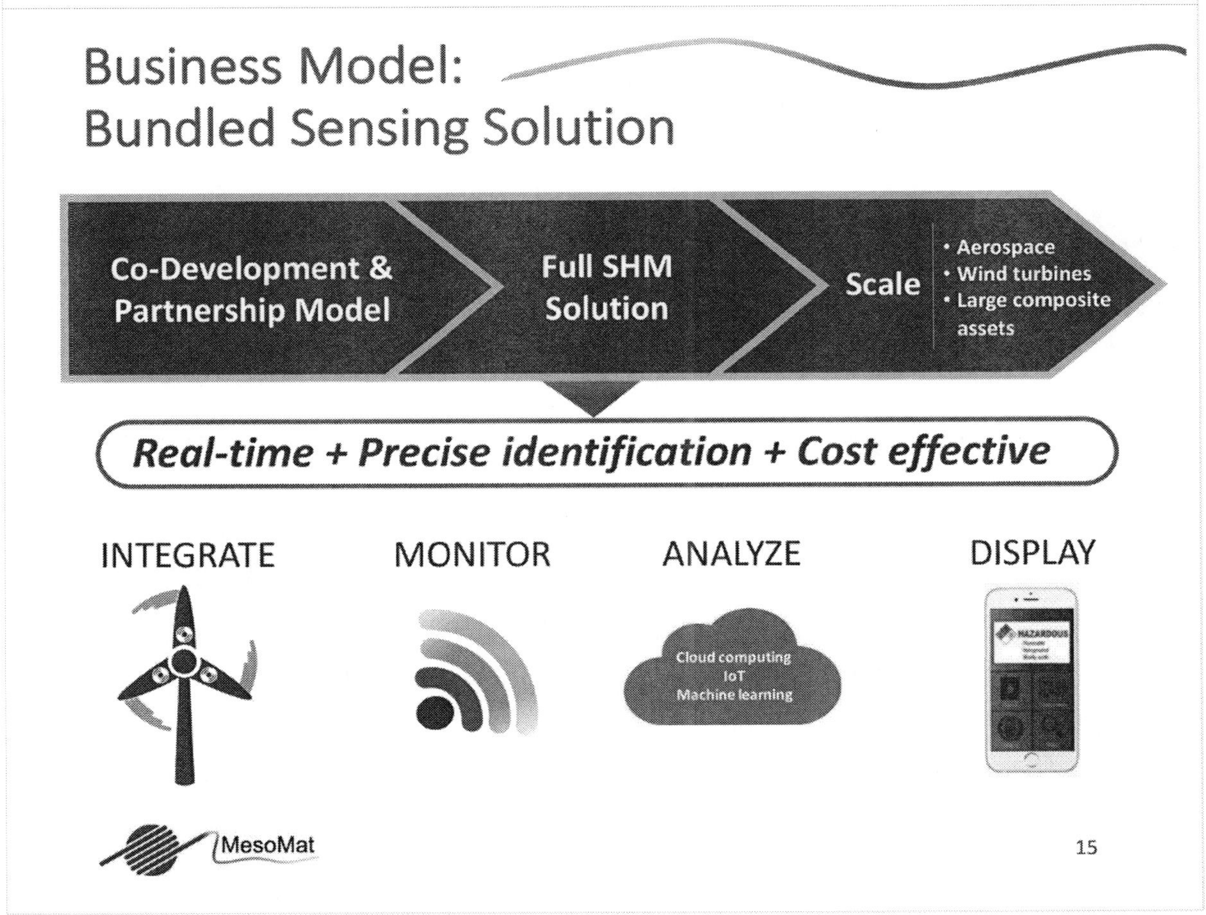
Business Model:
Bundled Sensing Solution
Co-Development & Partnership Model
Full SHM Solution
Scale
• Aerospace
• Wind turbines
• Large composite assets
Real-time + Precise identification + Cost effective
INTEGRATE
MONITOR
ANALYZE
DISPLAY
Cloud computing
IoT
Machine learning
MesoMat
15

Business Model at Scale

Manufacturing:
Tunable fibers for smart material solutions

Or

Licensing:
Specific application use with a targeted end application

→ Enable intelligent materials that integrate into electronics, sensing and data platforms

MesoMat

16

Competitive Advantage

Customizable

- Simple, modular approach for fiber modification
- Custom design via tunable manufacturing

Cost-effective

- Seamless integration
- Low input costs
- Fast production methods

Scalable

- Capital-light scalable, manufacturing
- Platform tech with broad value-added applications

Develop advanced material solutions

- Pathways to generate comprehensive IP portfolio
- Value-added fibers, including biodegradable and energy harvesting fibers
- Worldwide PCT on coil technologies

MesoMat

17

Aequor®

Antibiofilm Antifouling

Green Chemicals

Target Markets: >$100 B

Founders
Marilyn Bruno, Ph.D., J.D. - CEO
Cynthia Burzell, Ph.D. - CSO

AEQUOR CHEMICALS

SUSTAINABLE SOURCING

STRATEGIC ADVANTAGES

NON-TOXIC DISPERSANTS

- Green, GRAS, GRASE

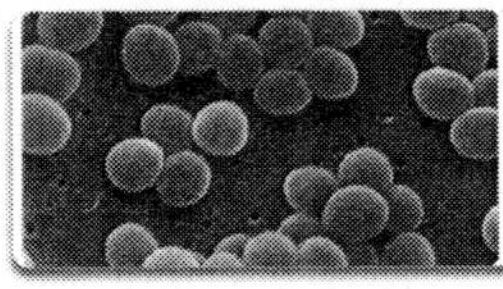

SUPERIOR ADDITIVES

- DISPERSE BIOFILM / FOULING
- ANTIMICROBIAL vs AMR strains
- SYNERGISTIC

CUSTOM PRODUCTS

- Surface cleaners (TRL-9)
- Water treatments (TRL-9)
- Personal hygiene (TRL-3)

Aequor
Antibiofilm Antifouling

THE TEAM

Cynthia Burzell, Ph.D.
CSO, Co-founder

Marine and Medical Microbiologist
COORDINATES WITH SAB, CONTRACTORS
RUNS OPS, VALIDATIONS

Marilyn Bruno, Ph.D., J.D.
CEO, Co-founder

Former Woman Entrepreneur of the Year – *Business Week*
COORDINATES VIRTUAL MODEL, RAISED > $1.2M

12 Proactive Advisors for Industry Insights
e.g.
Toshiyasu Nagae, Ph.D.
Former Pres, Dow-Japan

World Class Legal, Regulatory
e.g.
Richard Engler, Ph.D.
The Acta Group
K&L GATES

Aequor Syngene Certified SBA WOSB Woman Owned Small Business NASA USDA NIH

CHALLENGES ADDRESSED

REGULATORY PUSH

CUSTOMER PULL

FINDING EFFECTIVE REPLACEMENTS

Aequor

VALIDATED TECHNOLOGY

PROPRIETARY CHEMICALS AND UNIQUE BLENDS

SUPERIOR PERFORMANCE

- ANTIBIOFILM
- ANTIMICROBIAL
- KILL SUPERBUGS

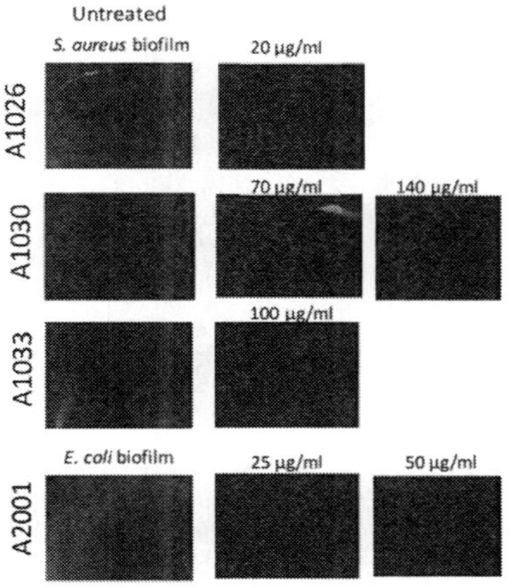

USDA

NASA

Aequor

Antibiofilm Antifouling

EXAMPLE: $50M PRESERVATIVE MARKET

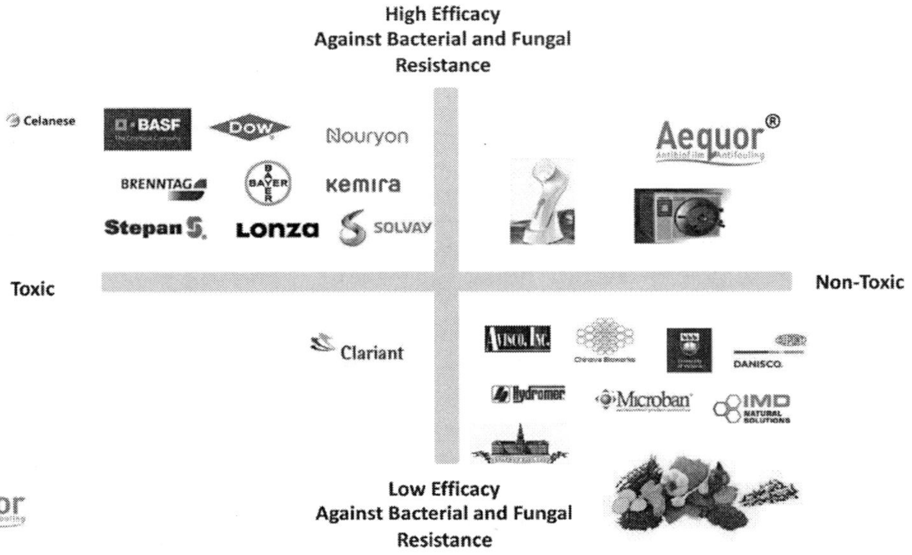

Aequor

Antibiofilm Antifouling

SEEKING $10M

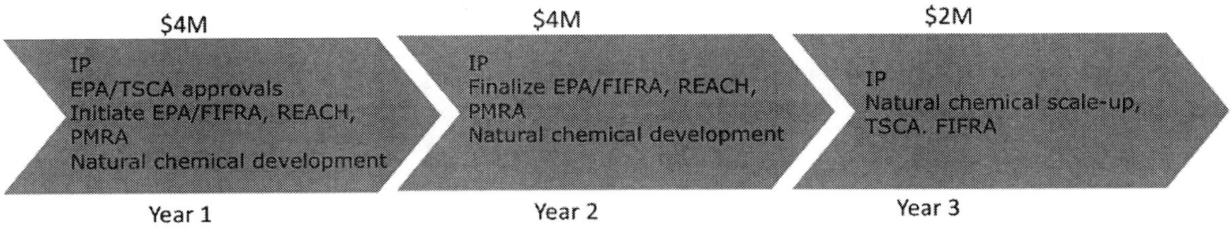

DELIVERABLES

- Scale-up of novel natural chemicals
- EPA/TSCA-approval for >25 additives
- EPA/FIFRA, REACH, PMRA approval for >20 additives
- IP – US, PCT, national stage in select international markets
- Revenues from consulting: assisting companies to Go Green
  B2B sales of custom products, including to U.S. government buyers

Aequor®
Antibiofilm Antifouling

- **OPPORTUNITY**
- **STRONG TEAM:** Expertise in Biofilm, AMR
- **SUSTAINABLE TECHNOLOGY ADVANTAGES and IP:** Additives, Synergy, Antibiofilm
- **SIGNIFICANT TRACTION**
- **SEEKING $10M**

www.aequorinc.com

Aequor

John P. Griffin CEO
j.griffin@terracoh-age.com

1

SOLUTION

TerraCOH

TerraCOH's proprietary technology enables zero-carbon geothermal energy production and energy storage

Present Status – TRL4

Our Mission

- Expand the global use of geothermal energy
- Deliver low cost, Emission-free, baseload (24/7) dispatchable electricity and grid-scale energy storage
- Store vast quantities of CO_2 permanently and safely underground

2

MANAGEMENT TEAM

TerraCOH

Experienced management team and with global expertise as well as extensive industry, engineering, & sales backgrounds:

John P. Griffin, CEO, Director/Chairman of Board

- **30+ years** experience in leading companies and organizations to success; as CEO and General Manager, he has taken organizations from zero to $200+ million in revenue
- B.S. Mechanical Engineering & M.B.A.

Dr. Jimmy B. Randolph, Chief Technology Officer & Inventor, Director

- **10+ years** experience in geothermal industry; 7+ years in CPG™ technology, CO_2 sequestration, geophysical fluid, heat transfer and energy storage
- Education: B.A. in physics and math; Ph.D. in geophysics

John F. Dolan, Business Development, Officer

- **20+ years** experience in legal industry ; 15+ years in mergers and acquisitions, technology transfer, IP acquisition, private equity and IP litigation
- Partner: Fredrikson & Byron, P.A. (Twin Cities 3rd Largest Law Firm)
- Associate: Patterson & Keough, P.A. (IP Prosecution and Litigation Law Firm)
- Patent Examiner: United States Patent and Trademark Office:
- Education: B.S in chemical engineering , B.S. in business administration, JD

3

PROBLEM

TerraCOH

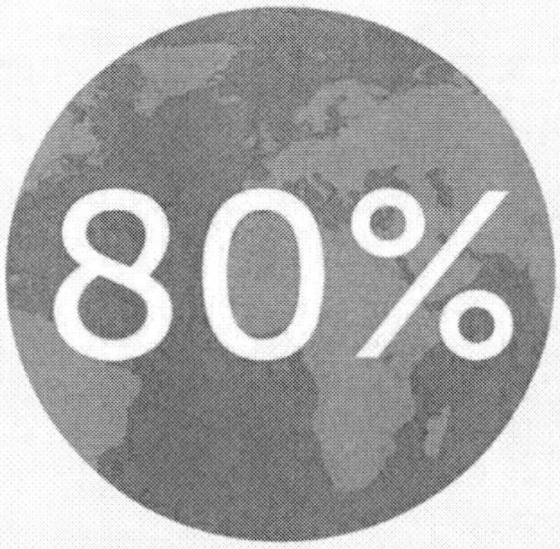

of the world's electrical power is generated from **fossil fuels**

4

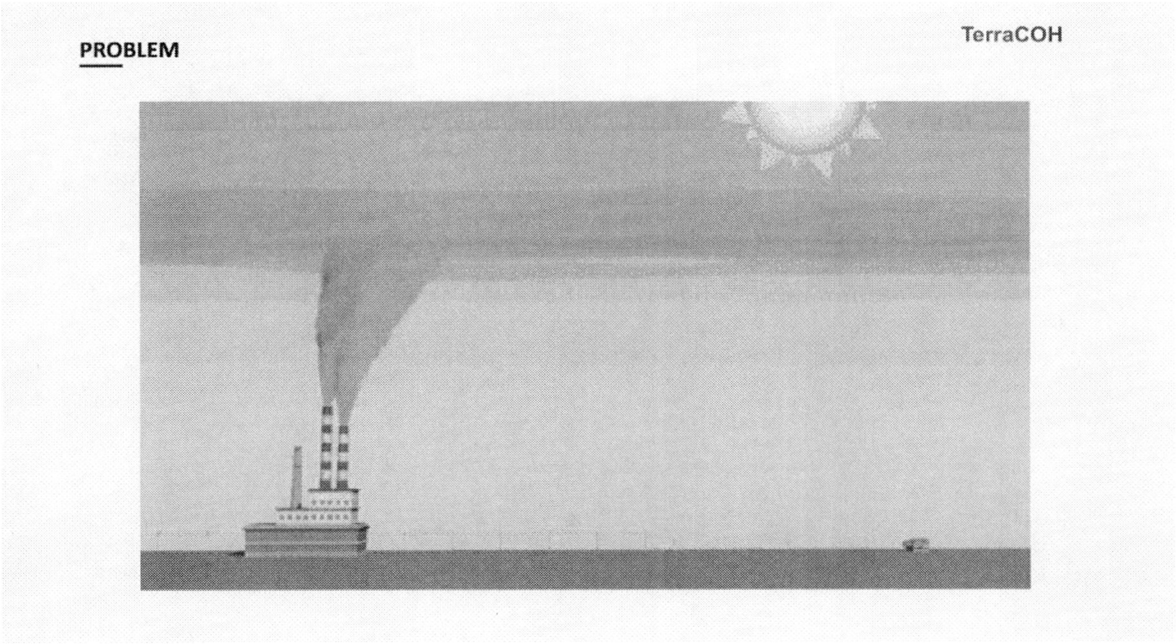
PROBLEM
TerraCOH

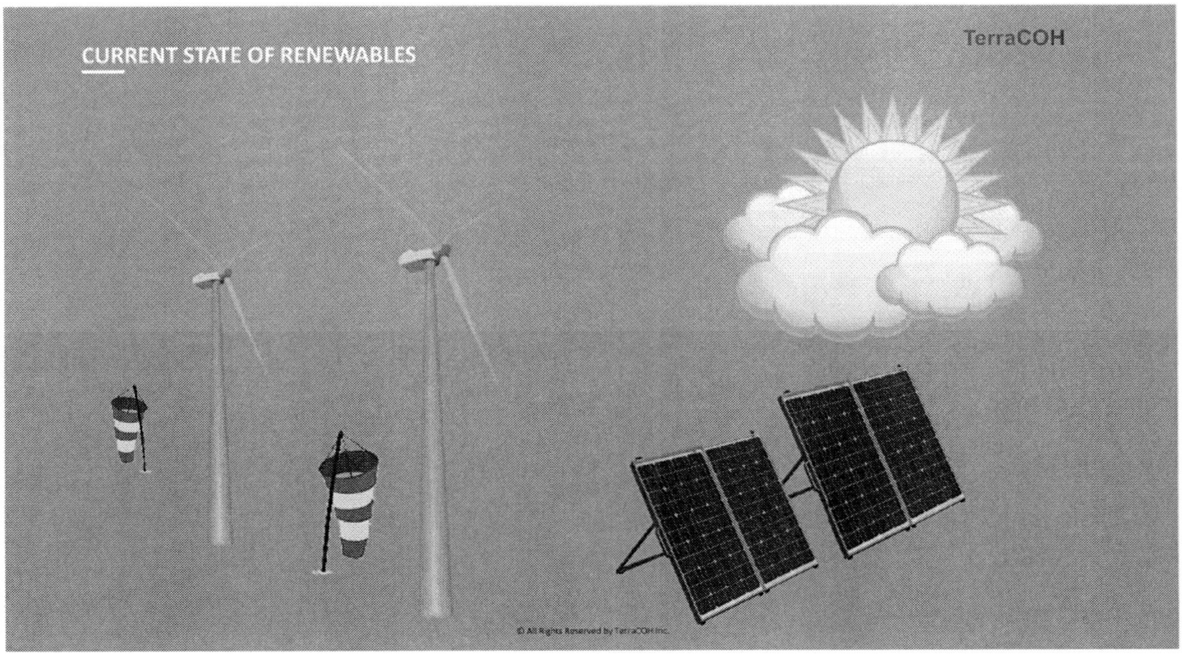
CURRENT STATE OF RENEWABLES
TerraCOH
© All Rights Reserved by TerraCOH Inc.

TerraCOH

Is there a Renewable Alternative?

7

GEOTHERMAL POWER

TerraCOH

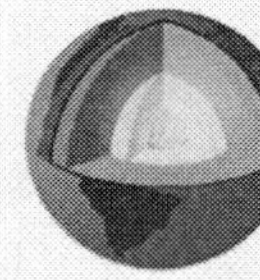

Geothermal Energy is Comparable to Fossil Fuels

- **Baseload** / Runs 24/7
- **Dispatchable** / Can turn On and Off on Demand

8

TerraCOH

SCALABILITY

Legacy Geothermal

Versus

TerraCOH

Expanding Geothermal Opportunities to more than **45%** of North America

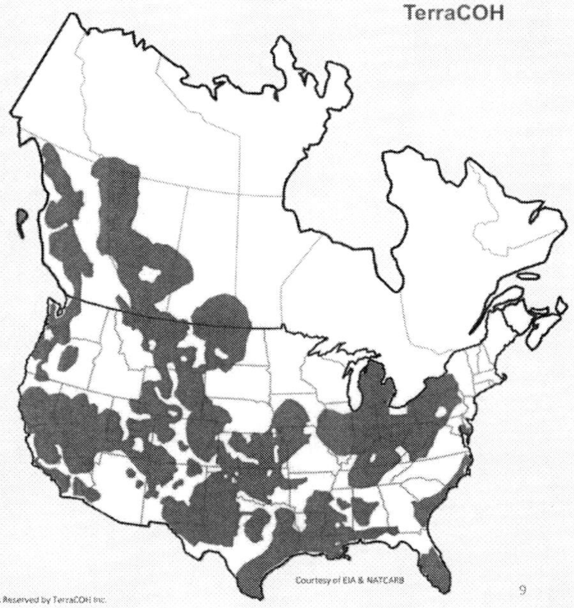

Courtesy of EIA & NATCARB

9

TerraCOH

SURPASSING FOSSIL FUELS

TerraCOH's Solution

- Cost – Lowest LCOE
- Zero-Emissions
- Consumes Tonnes of CO2
- Energy Storage
- Positive Legislation– 45Q

10

COMPETITVE LANDSCAPE

TerraCOH

LCOE
Levelized Cost of Electricity

*Cost of Power by Type (¢/kWh)

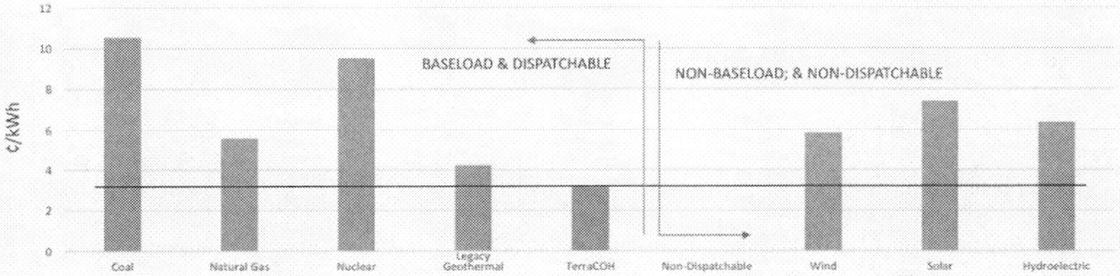

* Cost numbers from US DOE's Energy Information Administration (EIA), 2018 and estimated for TerraCOH

ZERO-EMISSION POWER

TerraCOH

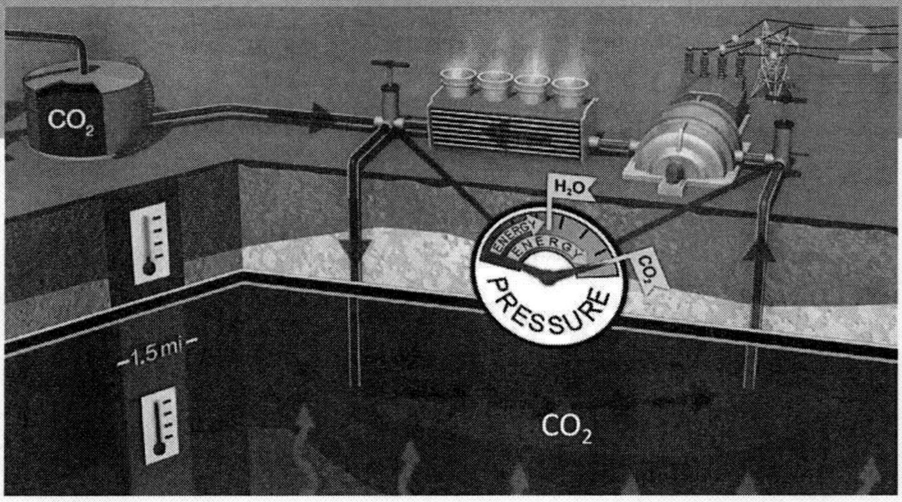

EARTH BATTERY ™

TerraCOH

Energy is stored as heat and pressure in the deep subsurface using CO_2 and brine.

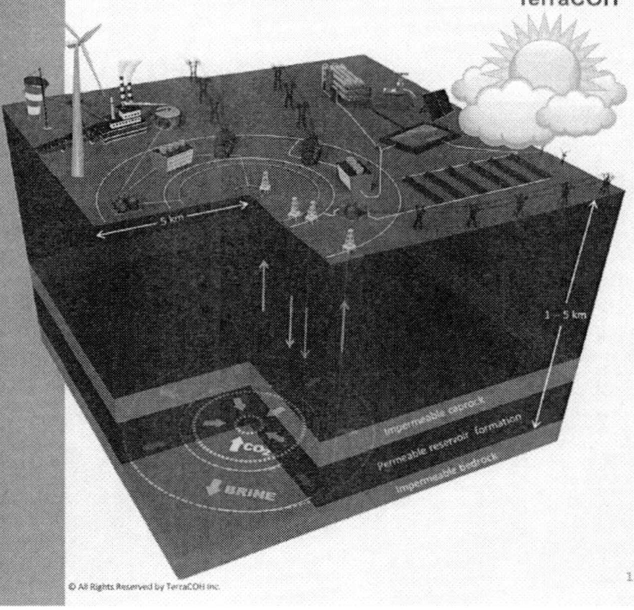

INTELLECTUAL PROPERTY

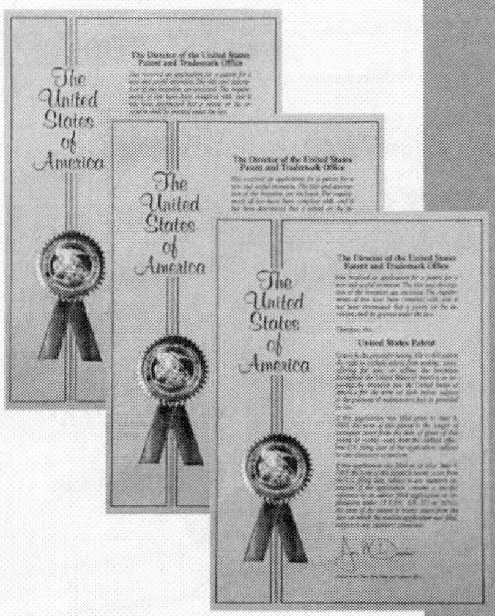

TerraCOH

Broad IP Portfolio

- **19 issued patents**
  (6 US and targeted countries worldwide)
- **8 pending**
  (2 US and targeted countries). E.g., U.S. Patent Nos. 8,991,510, 8,833,475 and 8,316,955

- IP Portfolio grants TerraCOH exclusive rights to the use of **"non-water based working fluids" (e.g. CO_2 and other fluids)** below ground for the generation of power and for energy storage
- Unique IP position that **covers all three phases** of TerraCOH's technology implementation plan

TARGET SEGMENTS

TerraCOH

- Industrial Emitters
- Oil & Gas
- Coal
- Biofuels
- Wind
- Solar

PARTNERS

TerraCOH

ECHOGEN power systems

SIEMENS

NSF

OHIO STATE

SBIR · STTR America's Seed Fund

ETH zürich

CLEAN TECH OPEN

+ NREL

Lawrence Livermore National Laboratory

ENVIRO Innovate

nrg cosia CARBON XPRIZE

ENVIRO Ambient

TerraCOH

BUSINESS MODEL

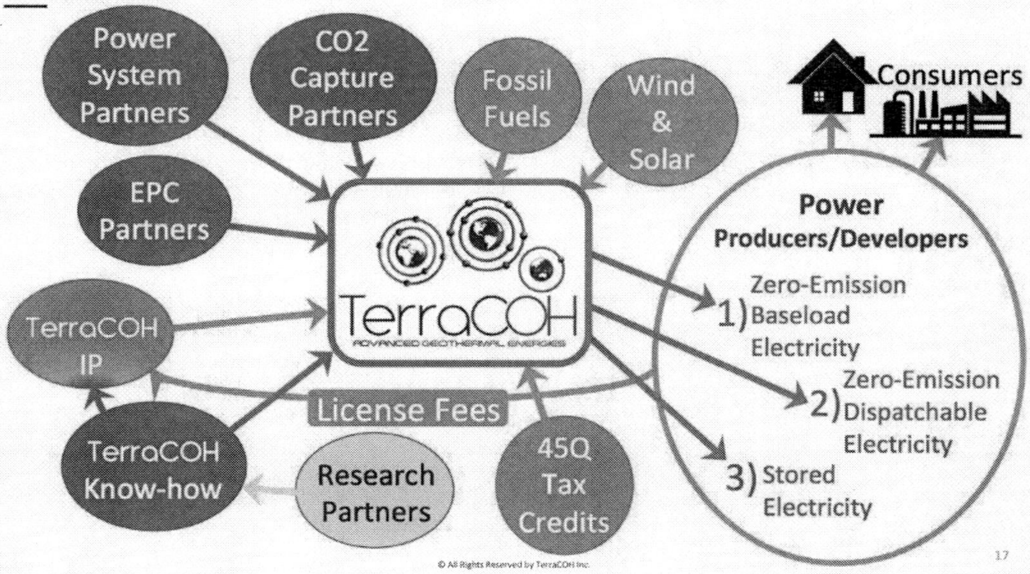

17

TerraCOH

DELIVERABLES / FUNDS REQUIRED

Path to MVP - $10-12 Million - 24-36 months

- At a "controlled" site environment; build a test bed to test various configurations (pressure, temperature, etc)
- Ideally partner with large CO2 emitting entity for the CO2
- Experience the permitting process
- Build a non-water based reservoir (CO2 or Nitrogen)
- Install small CO2 power system of about 1 MW
- Testing/verification of power system at field site
- Integrated heat source on site

18

MILESTONES

TerraCOH

TerraCOH's Journey

TerraCOH's proprietary technologies have and continue to gain momentum and Industrial/Societal accolades.

- Patents Filed and Granted
- Team TerraCOH formed 2014
- Licensed technology from UMN
- TerraCOH broadened Patents with additional Patents
- Collaboration with LLNL
- NRG Cosia XPRIZE – "Semi-Finalist"
- Additional Patents Granted
- Investment
- CCS and Industrial Partners
- Won DOE-SBIR Grant
- Cleantech Open Midwest Cohort – "Honorable Mention"
- Investment
- Shell Game Changer / NREL

TO DATE

- Additional Industrial connections
- Pilot
- Proof of Concept
- Commercialization of Technology

19

THANKS!

TerraCOH

ADVANCED GEOTHERMAL ENERGIES

Questions? Want to get involved?

Email: j.griffin@terracoh-age.com
www.terracoh-age.com
Follow us on Twitter - @TerraCOH_AGE

20

STF Technologies LLC

Richard Dombrowski

Co-Founder

18 Shea Way, Suite 101-102
Newark, DE 19713
Ph. 302-318-2027
E-mail: rddombrowski@stf-technologies.com

Our Business and Products

Motion Control Materials

STF-Armor™ - Flexible Protective Textiles

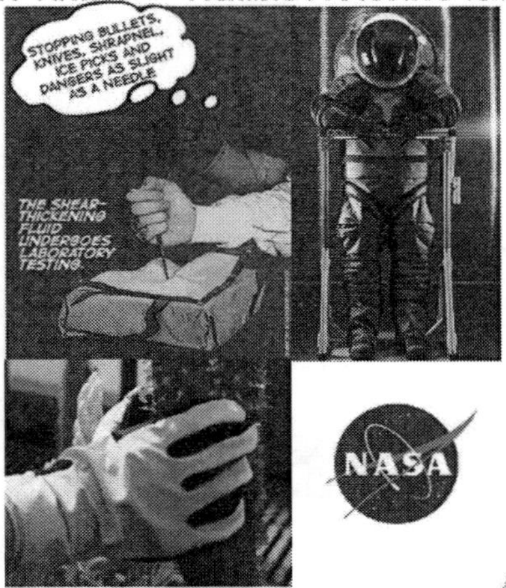

Image Credits: Reebok, US Army, NASA, US Dept. of State

Management Team

- **Richard Dombrowski** –CEO/Co-Founder – Full Time
  - Ph.D. Chemical Engineering
  - 10+ years STF-related research at UD & STF Technologies
  - 6+ years overseeing operations at STF Technologies
- **Norman Wagner** – Co-Founder – Part Time
  - Ph.D. Chemical Engineering, U. Delaware Chaired Professor
  - Co-inventor of STF-Armor™ and STF expert
- Engineering team includes 2 additional full-time chemical engineers

Our Goals:
Apply world-leading technical subject expertise to create materials and products that improve safety and health. Execute business plans and engage partners to achieve consistent, sustainable growth and profitability.

Chemical Ventures Conference – April 23, 2019 STF Technologies

The Challenge

How to solve conflicting functional materials requirements?

Personal Protective Equipment
Thin, Flexible, and Protective

Composites
Stiff and Tough

Prosthetics
Walking and Running

Athletic Apparel
Soft and Strong

And more...
Drilling Fluid, Lubricant, Automotive, Military...

Chemical Ventures Conference – April 23, 2019 STF Technologies

Shear Thickening Fluid: Patented, Proven, Scalable

STF: A Smart, Rate-Responsive "Liquid Ceramic"

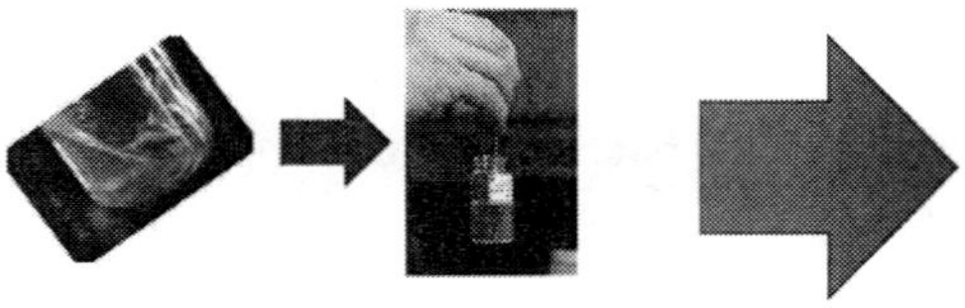

STF-Armor™ Textiles for Thin, Flexible, Multi-Threat Protection

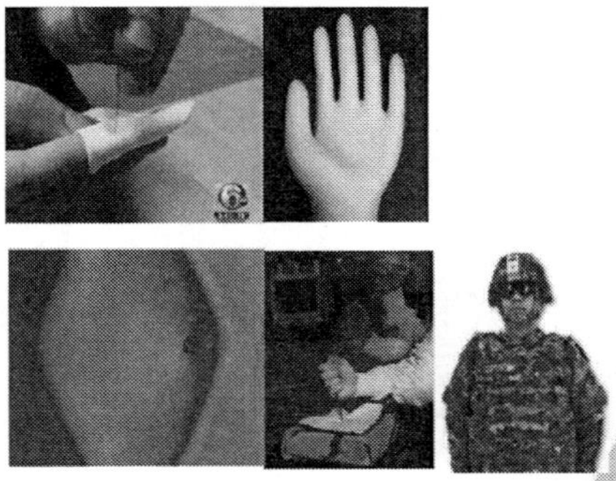

- Full commercial scale production of STF and STF-textiles
- Patents: US 7,226,878, 7,498,276 B2, 7,825,045, 10,244,801.
  - Other patents pending related to impact-resistant composites, aerospace

Image Credits: PEO Soldier - US Army

Markets and Competition

Market Size

- Personal Protective Equipment - $65 billion, CAGR 6%
- Surgical gloves - $1.1 billion, CAGR 8%
- Industrial safety gloves - $9 billion, CAGR 13%
- Composites (Impact Resistant) - $100 billion, CAGR 8%
- Prosthetics - $3 billion, CAGR 5%

Benefits to Customer

- Solves major pain points by resolving conflicting requirements
  - PPE - Major cost savings achievable through injury prevention
- Increased performance/efficiency, often with reduction in weight or thickness

Competition

- Conventional protective materials: HexArmor™, Turtleskin™ – stiff and uncomfortable
- Dynamic systems (prosthetics, motion control): sensors & electronics – batteries, weight

We are seeking…

- Licensing and Development Partners
  - Markets aligned with existing development efforts - protective materials, PPE, surgical gloves, first responder gear, composites
  - Product and materials problems requiring a "dual natured" or smart rate responsive behavior/solution
- Investor Partners
  - Large single market and/or platform opportunities

Industries

Aerospace
Automotive
Composites
Consumer products/apparel
Cushioning and packaging
Industrial personal protective equipment
Lubricants and power transmission fluids
Medical devices – prosthetics
Military – vehicle and body armor
Sporting goods

Thank you!

For inquiries, please contact:

Richard Dombrowski
Co-Founder
Ph. 302-318-2027
E-mail: rddombrowski@stf-technologies.com

STF Technologies LLC
18 Shea Way, Suite 101-102
Newark, DE 19713

PLIANT
POWER DEVICES

Electric Plastics for next generation batteries

Today Batteries Are...

Expensive

Heavy

Unsafe

PLIANT
POWER DEVICES

Metal-based materials have dominated for 30 years.

Pliant is changing the game with ***Electric Plastics***.

PLIANT POWER DEVICES

Solution - **Electric Plastics**

~50% cost of ownership

Safe to >200C

>15% lighter

Compatible with existing manufacturing

PLIANT POWER DEVICES

Copyright © Pliant Power Devices Inc. 2019

Technology

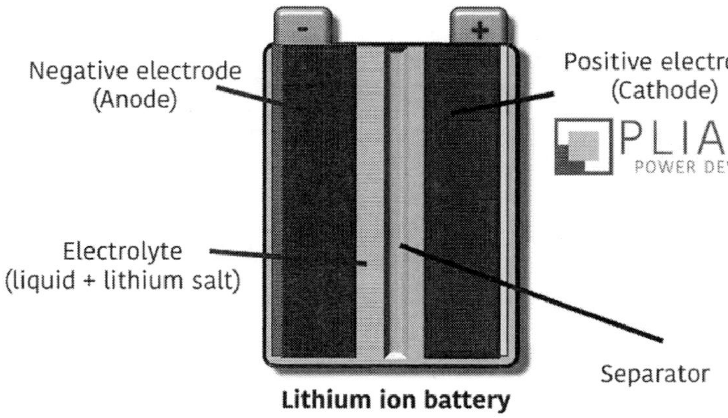

Lithium ion battery

3 Patents and Applications

11 Peer-Reviewed Publications

PLIANT POWER DEVICES

Market Opportunity

Lithium Ion Cathode Market 2023: 11B USD

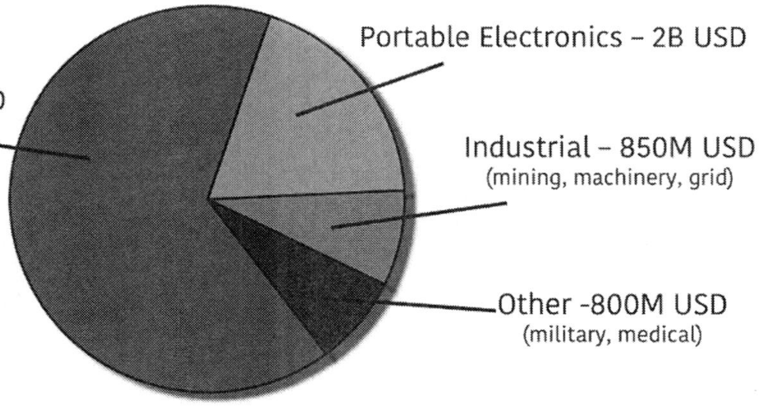

PLIANT POWER DEVICES

Source: MarketsandMarkets. 'Battery Materials Market by Battery Type.' (2018)

Competitive Landscape

Company (material)	Cell Energy (Graphite Anode)	Cell Energy (Lithium Anode*)	Cycle Life	Safety
PLIANT POWER DEVICES (PCM*)	**300 Wh/kg**	**550 Wh/kg**	**Moderate -High**	**Excellent**
umicore, NICHIA (NMC)	260 Wh/kg	450 Wh/kg	Moderate	Moderate
JM Johnson Matthey, A123 SYSTEMS, BYD (LFP)	125 Wh/kg	190 Wh/kg	High	High

PCM = Pliant Cathode Material, NMC = Nickel Manganese Cobalt, LFP = Lithium Iron Phosphate

PLIANT POWER DEVICES * Expected values

Key Team Members

Management Team

Dr. Tyler Schon
CEO

Dr. Grayson Ingram
CTO

Supporting Team

Bryony McAllister
Materials R&D

Soyoung An
Materials R&D

Irene Lau
R&D Engineer

Scientific Advisor

Prof. Dwight Seferos
UofT Full Professor

PLIANT POWER DEVICES

Financial Projections

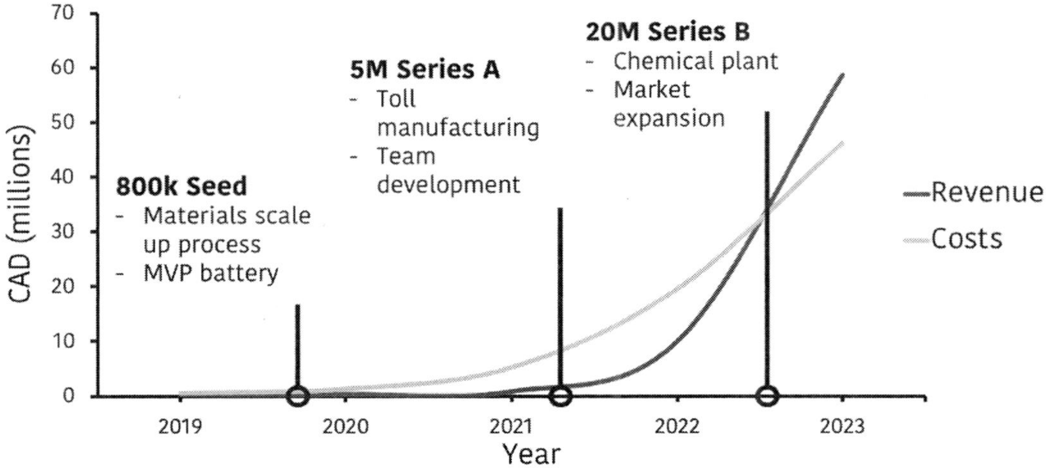

 Values in CAD

PLIANT POWER DEVICES

Electric Plastics for next generation batteries

Tyler Schon
tyler.schon@pliantpower.com

Making CO_2 Green

Apoorv Sinha- ChBE- Georgia Tech ('10)
a.sinha@carbonupcycling.com
1-403-400-7690
@CarbonUpcyclin
g

1

Products

Portfolio of fine additives:

GNPs	EFA	GO	rGO	GQDs
Graphitic Nanoplatelets	Enhanced Fly Ash	Graphene Oxide	Reduced Graphene Oxide	Graphene Quantum Dots

All derived from CO_2 & cheap solids

Management Team

Name	Title	Experience
Apoorv Sinha	Project Lead, CEO, Chem Eng	Chem Eng; +7 years startups & advanced materials
Greg Boser	Director	+35 years energy sector and hardware startups
Robert Foster	Mechanical Lead, P. Eng	Mech Eng; +35 years high pressure equipment, waste water
Bob Kramer	Market Lead, Concrete & Coatings	+35 years in concrete industry
Jeremy Olson	Controls Lead, P. Eng	Elec. Eng; +15 years in controls design and automation
Abraham Kannanat	Chemistry Lead, PhD	PhD Chemistry; +35 years' organic synthesis and nanomaterials
Naguib/Birss/Peterson/Pu Chen/ Ma/Siva	Academic Partners	100+ years' combined expertise in concrete, polymers, composites, nanomaterials, solar cells, and pharmaceuticals

3

@CarbonUpcycling

Technology

Chemical Adsorption

Catalytic Process

Graphite

Fly Ash

Patent-pending

10-30%

CO_2 content

200° C

Temp Stability

4

@CarbonUpcycling

Products & Markets

Carbon Upcycling

Coating	Concrete	Polymers	Adhesives
			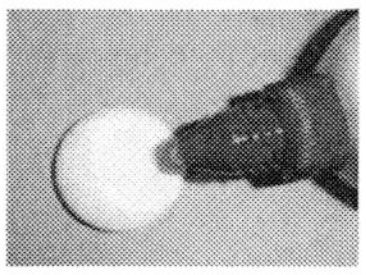
Commercial Product Concrete & steel substrates	Improves fly ash quality Reduced cement consumption	PE Composites tested Nylon, PET, PU work in progress	PU Adhesives tested Lignin composite work in progress
CRH, Florida Lemark	LafargeHolcim BURNCO		

5

@CarbonUpcycling

Status & Financing

Carbon Upcycling

Market applications identified in 10+ areas

PCT Patent Filed

nrg | cosia CARBON XPRIZE

Funds raised to date- USD 7 MM

Support from most major Canadian institutions- National Research Council, Alberta Innovates, NSERC, Natural Resources Canada, Emissions Reduction Alberta

Current Status- 10 TPD unit in Calgary, AB

6

@CarbonUpcycling

Interest

Identify market or development partners to:

- Incorporate CUT CO_2-derived nanomaterials
- Show Circular Economy in market
- Find tangible product lines for deployment of materials in marketplace

Engage Further Strategic Partners Active in

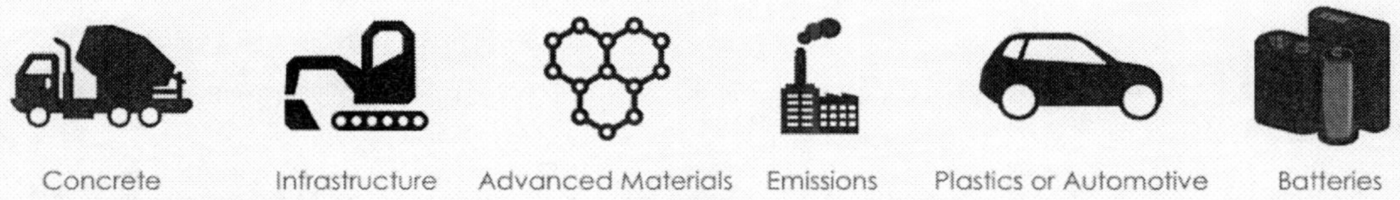

Apoorv Sinha- CEO a.sinha@carbonupcycling.com 1-403-400-7690

AlphaCarbon Coating

CUT's 1st Commercial Product (April '17)

Corrosion-coating for concrete

Cost-competitive

Currently used in OR & FL

Enhanced Fly Ash

- Derived from Fly Ash
- Compressive Strength (>30%)
- Trials with 5+ companies
- Cost-competitive
- CO2 content/kg: >15%
- Trials with Lafarge, Argos, & Heidelberg

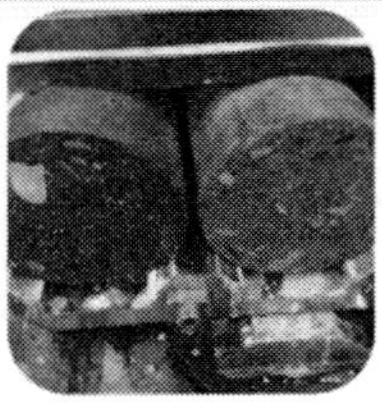

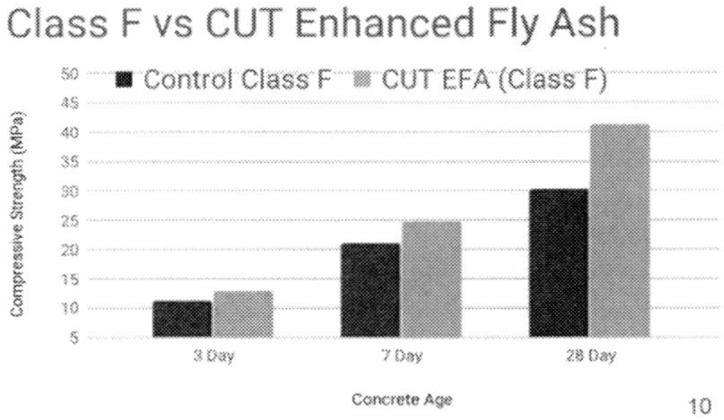

10

@CarbonUpcycling

Concrete Coatings

Global Concrete Corrosion Market
$3,27 BN Annual Market

OldCastle Manholes
$14M Annual Market

OldCastle Grease Traps
$1.4 M Annual Market

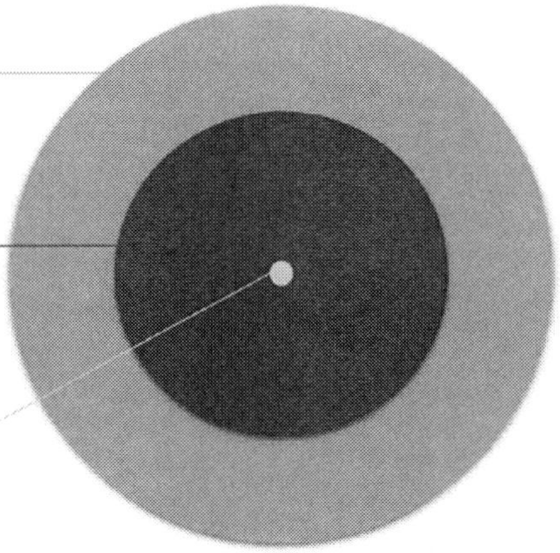

11

@CarbonUpcycling

Value Proposition

Carbon Upcycling

Current Concrete Makeup

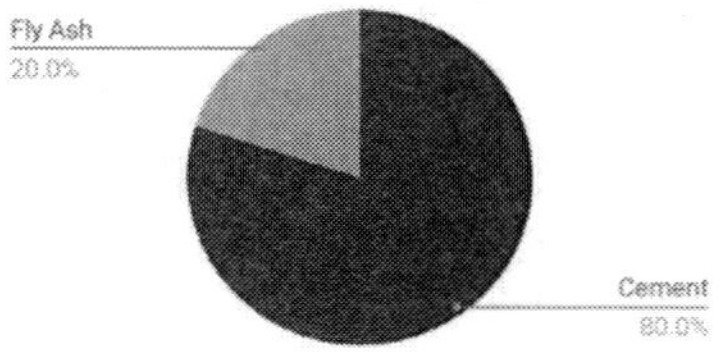

Material Prices

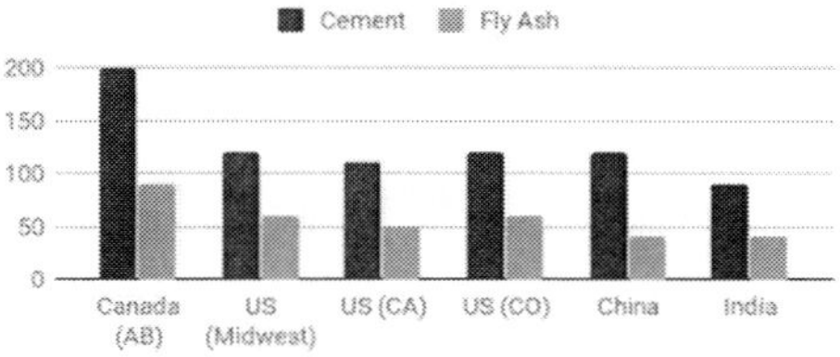

Inconsistent Quality

Lower Early Strength

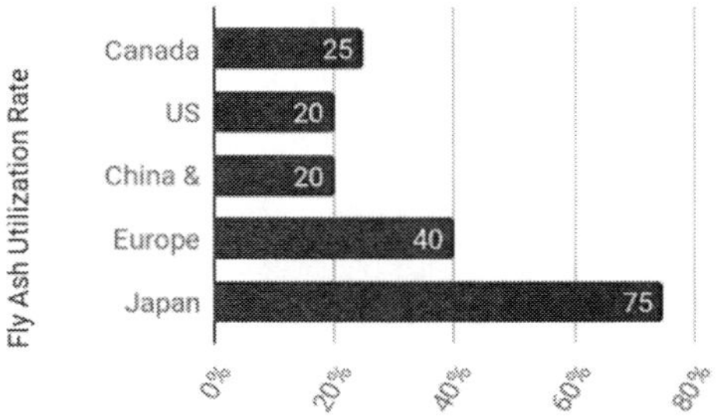

Potential Mix Design (w/ CUT EFA)

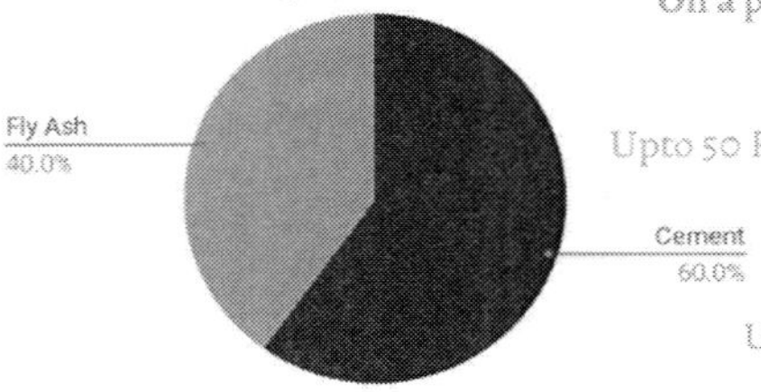

On a per cubic meter concrete basis:

Upto 50 Kg CO2e reduction (25%)

20% Cement Reduction

Upto 10% Cost Reduction

12

@CarbonUpcycling

Fly Ash Market

Carbon Upcycling

Global Fly Ash Market
$5.97 BN Annual Market

US Fly Ash Market
$3.35 BN Annual Market

Alberta Fly Ash Market
$200 M Annual Market

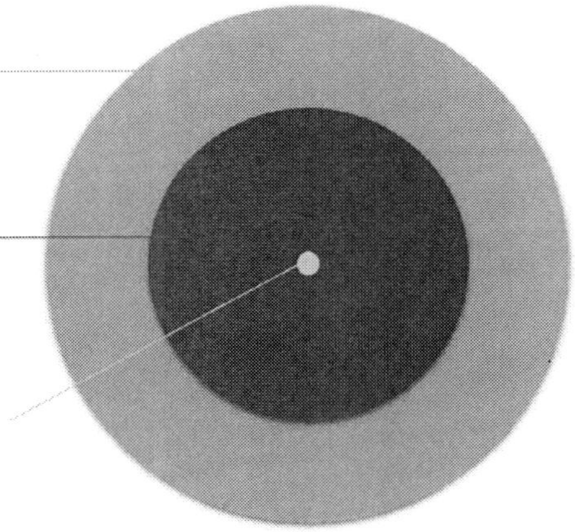

Co_2 Reduction/Kg EFA used: 1-1.5 Kg CO2e

13

@CarbonUpcycling

Potential Impact

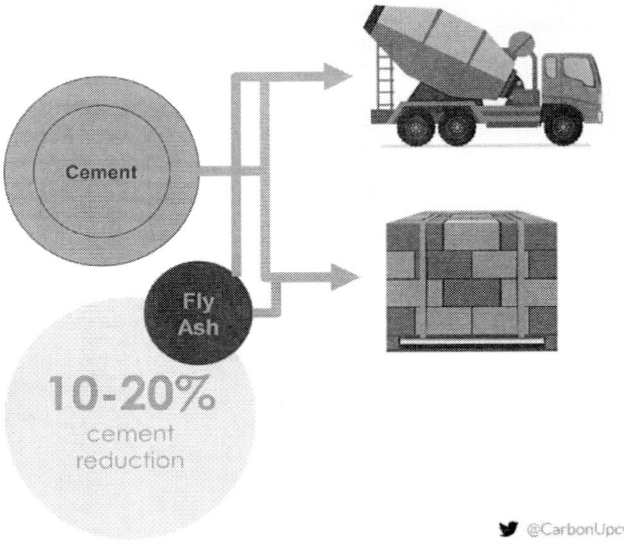

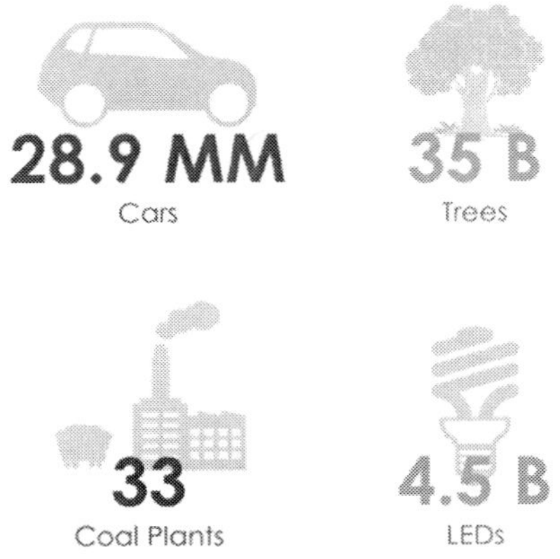

14

@CarbonUpcycling

Other Markets

Water Filtration Membranes

Asphalt Additives

Solar Cell Coatings

Rubber Additives

Identified through technical validation from accredited Universities

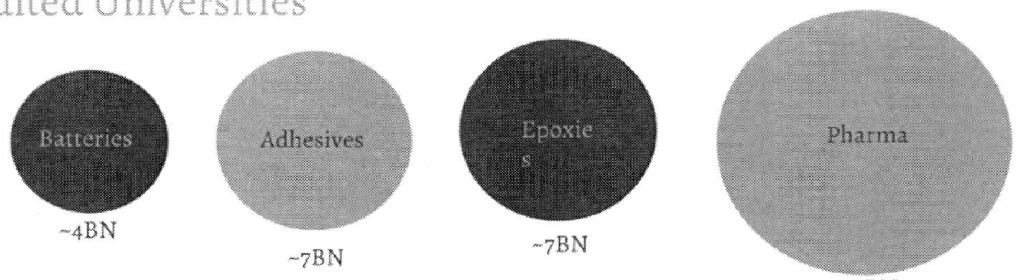

	Batteries	Adhesives	Epoxies	Pharma
Global Market	~4BN	~7BN	~7BN	>10BN
CO_2 Reduction/Kg Used:	~1-1.5	~5-85	~5-85	~N/A

15

@CarbonUpcycling

Business Model:

B2B Sales of additives and cementitious materials-

Concrete Coating AC-100	Enhanced Fly Ash EFA-100	Plastic Nucleator PEN-100
Direct Sales	Ready-mix and precast concrete companies	Plastic resin manufacturers
Municipal and city specifiers	Licensing or joint development projects	Part manufacturers (automotive, retail, packaging, etc)

@CarbonUpcycling 16

Funding <2018

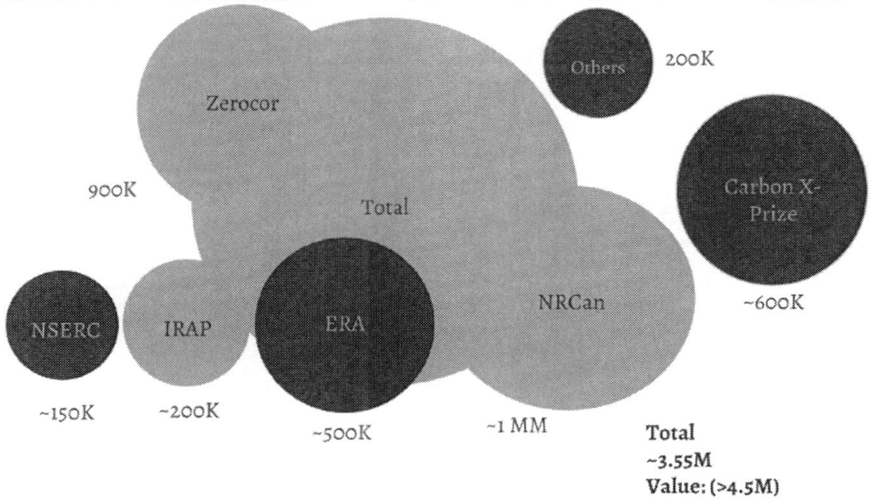

@CarbonUpcycling 17

Management Structure

Carbon Upcycling

Carbon Upcycling Technologies

- **VP Engineering & Design- Rob Foster**
  - Area 51 Engineering
  - Addos Controls
  - Southern Research
- **Operations Manager- Chad Belanger**
  - Jamie Merriam- Safety
- **Management & Strategy**
  - CEO- Apoorv Sinha
  - Director- G. Boser
  - BDLS Group
- **Business Development**
  - Concrete- B. Kramer
  - Plastics- N. Tucker
- **Admin**
- **Research & Development**
  - Chemist- A. Kannanat
  - Waterloo- P Chen
  - Calgary- V. Birss
  - Toronto- H. Naguib

18

@CarbonUpcycling

Ecosystem Support

Carbon Upcycling

UNIVERSITY OF CALGARY

ALBERTA INNOVATES

Universitat d'Alacant
Universidad de Alicante

SAPL

CTA CANADIAN TECHNOLOGY ACCELERATORS

UNIVERSITY OF Waterloo

KINETICA VENTURES

nrg | cosia CARBON XPRIZE

Mechanical & Industrial Engineering UNIVERSITY OF TORONTO

NSERC CRSNG

CREATIVE DESTRUCTION LAB

Natural Resources Canada / Ressources naturelles Canada

Canadä

UBC

NRC CNRC
NRC Industrial Research Assistance Program

EMISSIONS REDUCTION ALBERTA

BERKELEY LAB

19

@CarbonUpcycling

Environmental Benefits

CO2 Utilization (>300 Kg CO_2 per day)

Solid Waste Upcycling (9 TPD Fly Ash)

Waste heat utilization: Greenhouse

No water use

Circular Economy Demonstration

CO2-derived products

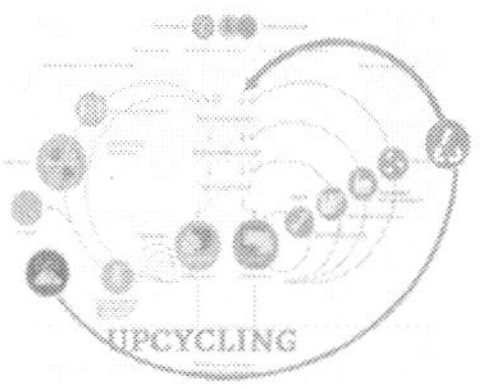

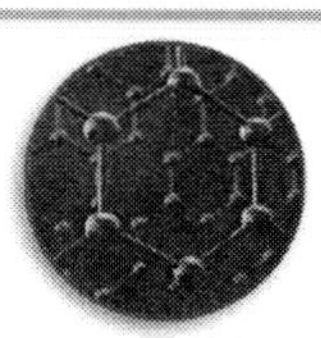

20

@CarbonUpcycling

Timeline & Milestones

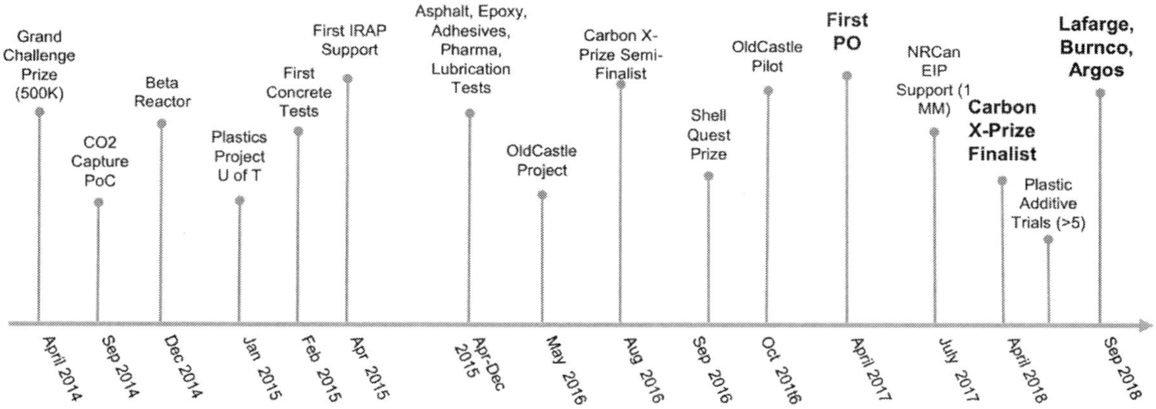

21

@CarbonUpcycling

Polymer Additive

- Polyethylene (PE)
- Mechanical Reinforcement
- Increase production throughput
- Increase product stiffness
- Engaged in commercial trials

@CarbonUpcycl

Polymer Additives

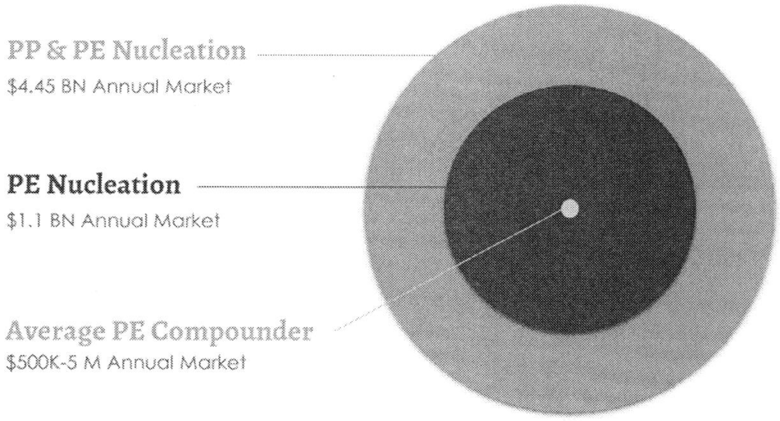

CO_2 Reduction/Kg GNP used: 5-85 Kg CO2e

23

@CarbonUpcycling

Competition

@CarbonUpcycling

24

Potential Impact

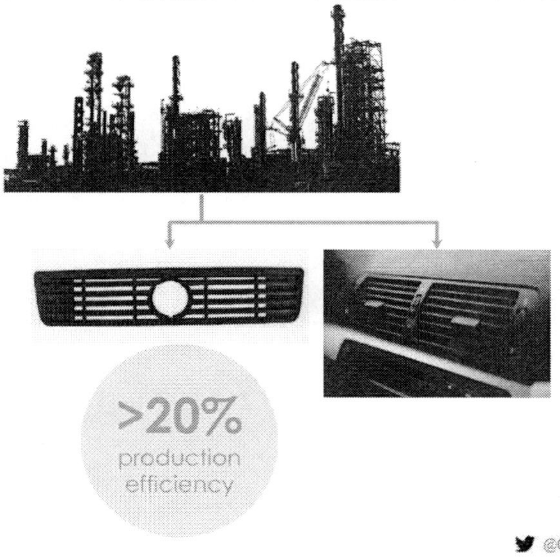

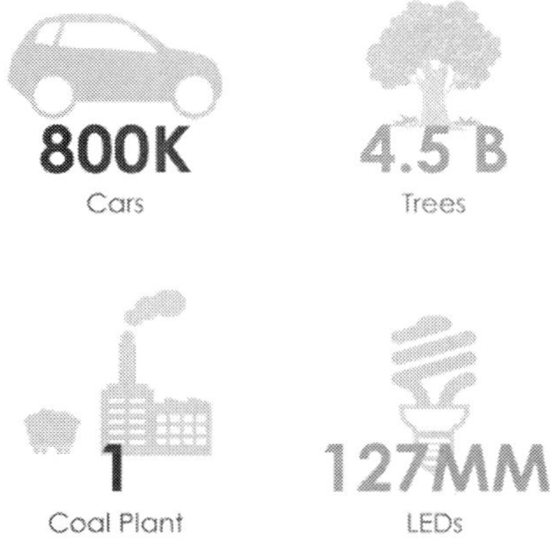

@CarbonUpcycling

25

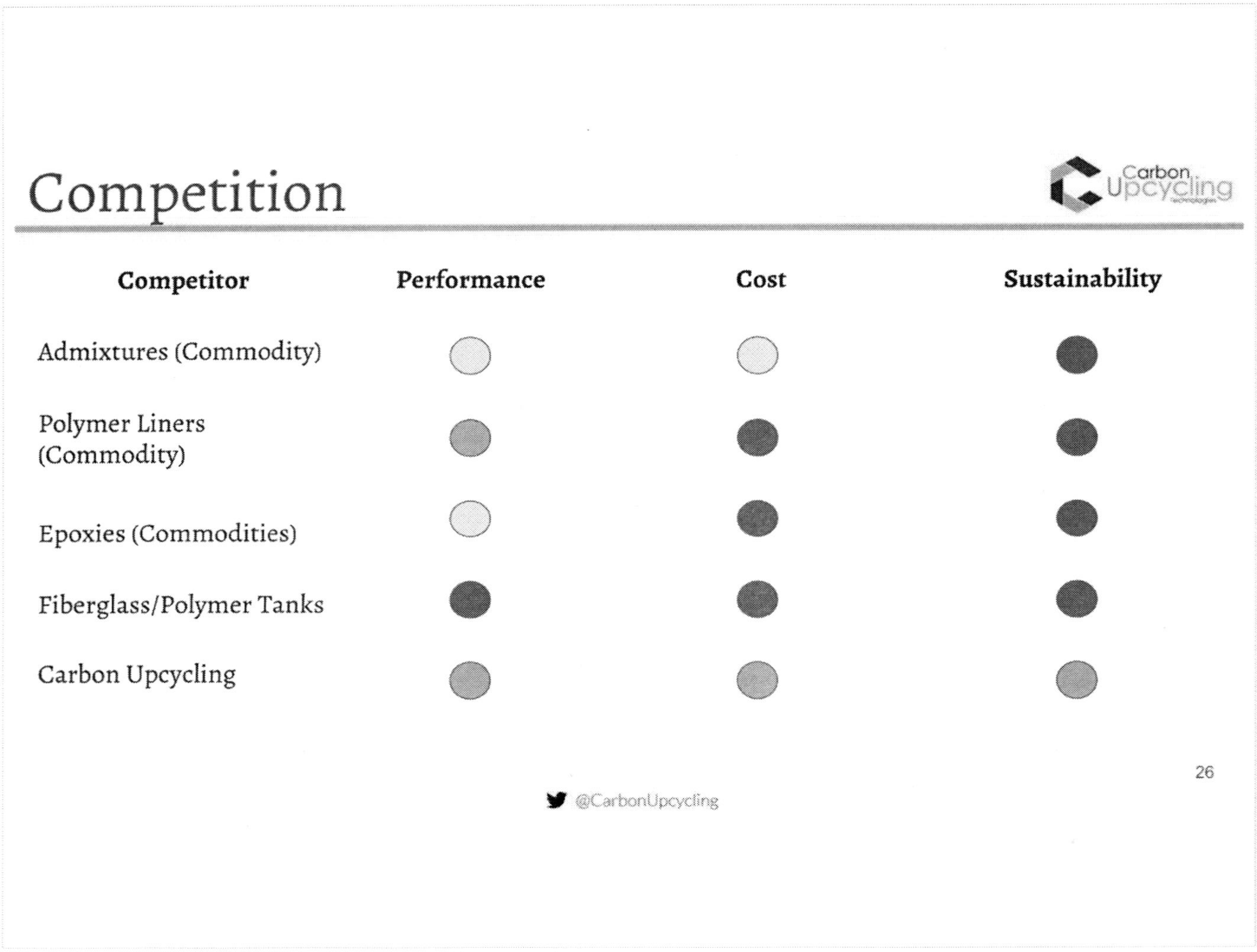
Competition
Carbon Upcycling
Competitor
Performance
Cost
Sustainability
Admixtures (Commodity)
Polymer Liners (Commodity)
Epoxies (Commodities)
Fiberglass/Polymer Tanks
Carbon Upcycling
26
@CarbonUpcycling

iMiCROBES

Commodity Chemicals from Methane

Derek Greenfield, PhD
Co-founder & CEO

What Can We Learn?

AMYRIS
opxbio good chemistry.
solazyme
Cobalt
LS9
Rennovia

vs.

Industrial Biotech Cannot Make Commodity Chemicals?

Methane is the Path Back to Commodity Chemicals

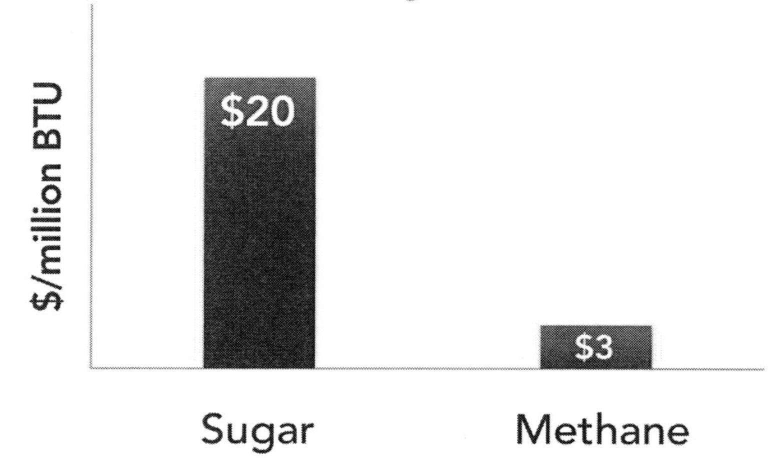

Methane From Waste

Commodity Chemicals from Methane

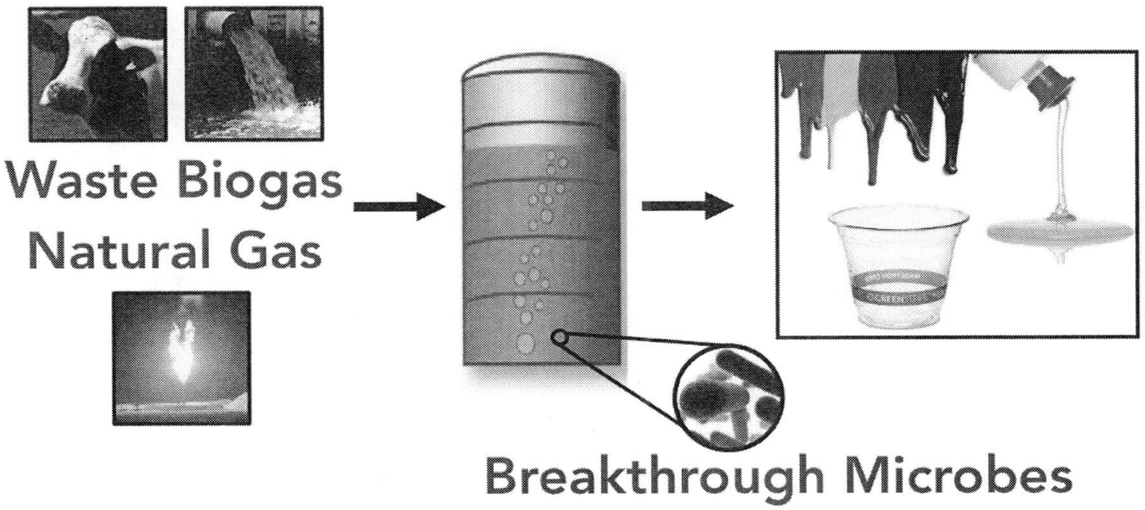

Our Technology: Programmable Methane Microbes

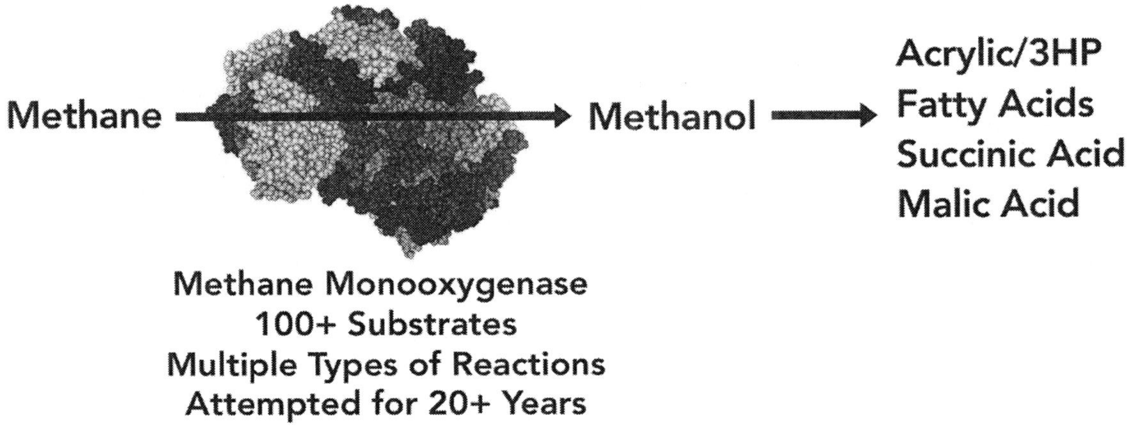

What Makes Us Different: Synthetic Biology Approach

Methanotrophs

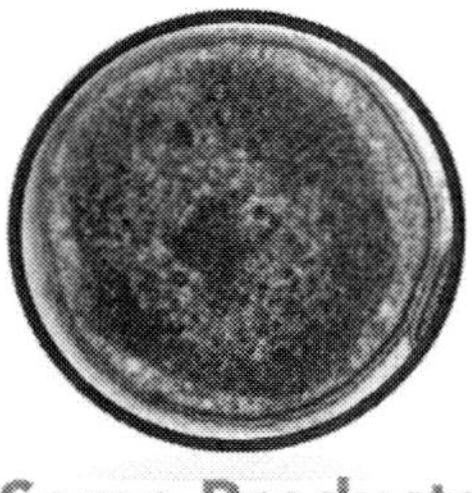

Some Products
Slow to Engineer

Industrial Microorganisms

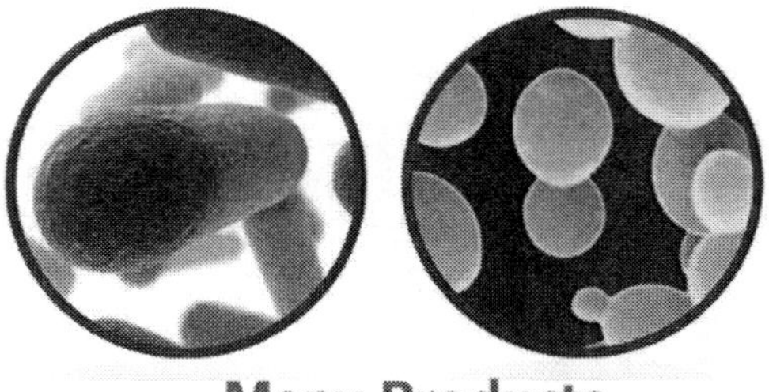

Many Products
Fast to Engineer

Progress to Date

Seed Funding

Y Combinator
SV TECH VENTURES
CRM
CLI VENTURES
MICROVENTURES

R&D Grants

arpa·e
Advanced Research Projects Agency • ENERGY
United States Environmental Protection Agency
ERA
Department of Energy
USDA
NSF

Founding Team

Dr. Elizabeth Clarke, CTO
Berkeley
PhD UCSF

Dr. Derek Greenfield, CEO
Harvard
PhD Berkeley

Dr. Noah Helman, CSO
Harvard
PhD Stanford

Passionate, Visionary Team with Commercialization Experience
Founders Have Worked Together For 10 Years

What We Have Learned

- World now is different than 10 years ago
- Tools for synthetic biology rapidly improving
- Excitement around new feedstocks for biology
- Methane is the path to bio commodity chemicals
- When others assume something is impossible...

iMiCROBES

Commodity Chemicals from Methane

Derek Greenfield, PhD
Co-founder & CEO

CLEAN TECH OPEN
2018 NATIONAL WINNER

Radical PLASTICS

a **radical** solution to **plastic** pollution

Kristin Taylor, Co-founder / CEO
ktaylor@radical-plastics.com
(617) 909-9032

Dr. Yelena Kann, Founder / CTO
ykann@radical-plastics.com
(617) 320-9663

www.Radical-Plastics.com

Radical PLASTICS

Company Overview

Radical PLASTICS

- **Patent-pending, catalyst technology that renders conventional plastics biodegradable in the natural environment**
- **Produced compounded resins and film at pilot scale** using standard processing equipment
- **Completed physical testing** that shows that RP compounds retain the mechanical properties of conventional plastics (LDPE)
- **Completed analytical testing** that shows that our proprietary, non-toxic, naturally occurring catalyst changes the chemical structure of the polymer, turning it into a low molecular weight polyethylene wax which is known to be readily biodegradable
- **Raising a seed round of $1.3M** with breakeven expected in 2021

Radical Plastics is providing a game-changing solution to plastic pollution

Plastics are AMAZING!!!

- Strong
- Lightweight
- Cost-effective
- Versatile
- $650B Industry

Plastics are AMAZING!!!
... but Plastics are AWFUL for the Environment

- 380 million tons/year
- 25% is packaging
- 12 B tons in landfills by 2050
- 9% is recycled (~4.4% since 2018)
- 8 M tons in our oceans

The Radical Plastics Solution

Fine mineral matter → Proprietary free radical catalyst + Conventional plastic resins → Radical PLASTICS

A nontoxic, natural byproduct of the mining industry

Catalyst, Co-catalysts & Inhibitors

Polyolefin Plastics: polyethylene (LDPE, LLDPE, HDPE), polypropylene

Plastic compounds that are FULLY BIODEGRADABLE in the natural environment

Radical Plastics will be selling a fully formulated biodegradable plastic compound

- Selling formulated plastic pellets to processors
- Control the onset of degradation (useful life)
- Control the rate of degradation (4 months-2 years)
- USPTO & PCT filing March 2019

Two Stage Degradation Process

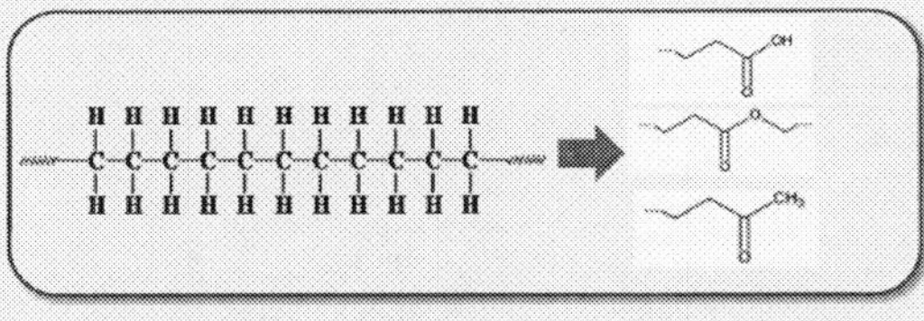

Biomass
CO_2
H_2O
Mineral nutrients

Chemical Phase

- Catalyst system breaks the long polymer chains into smaller components
- Inserts polar groups along the chain to make the polymer hydrophilic (wettable surface)
- Allows a "biofilm" to form on the surface – First stage of biodegradation
- Does not result in microplastics!

Biological Phase

- Microbes are found in the environment (soil, oceans, fresh water, etc.)
- Microbes metabolize the material and convert it into biomass, CO2 and water
- Trace amount of mineral nutrients that are healthy for the soil
- Meets EPA 503 regulations

Radical PLASTICS

Initial Target Market – Agricultural Mulch Film ($4B)

Mulch film…
Prevents Weeds
Decreases Water Usage
Increases Crop Yield

Conventional Plastic Film
Mulch film = $180/acre
Removal & Disposal = $250/acre
TOTAL = $430-520/acre

Mulch Film made from the Radical Plastics Compound
Mulch Film = $400-500/acre
BIODEGRADATION = $0
TOTAL =$400-500/acre

7

Radical PLASTICS

Competitive Landscape – Mulch Film Market

	Pros	Cons
Conventional plastic mulch film (LDPE)	√ Low Cost √ Excellent Properties √ Ease of Use	X $$$ to remove X High GHG emissions
Starch and starch/ polyester blends	√ Will biodegrade (starch)	X Poor physical properties X Difficult to use
Radical PLASTICS	√ Low cost of production √ High Margins √ Excellent Properties √ Fully Biodegradable	

Radical Plastics compound is a **drop-in replacement** for LDPE with the added benefit of complete biodegradation

8

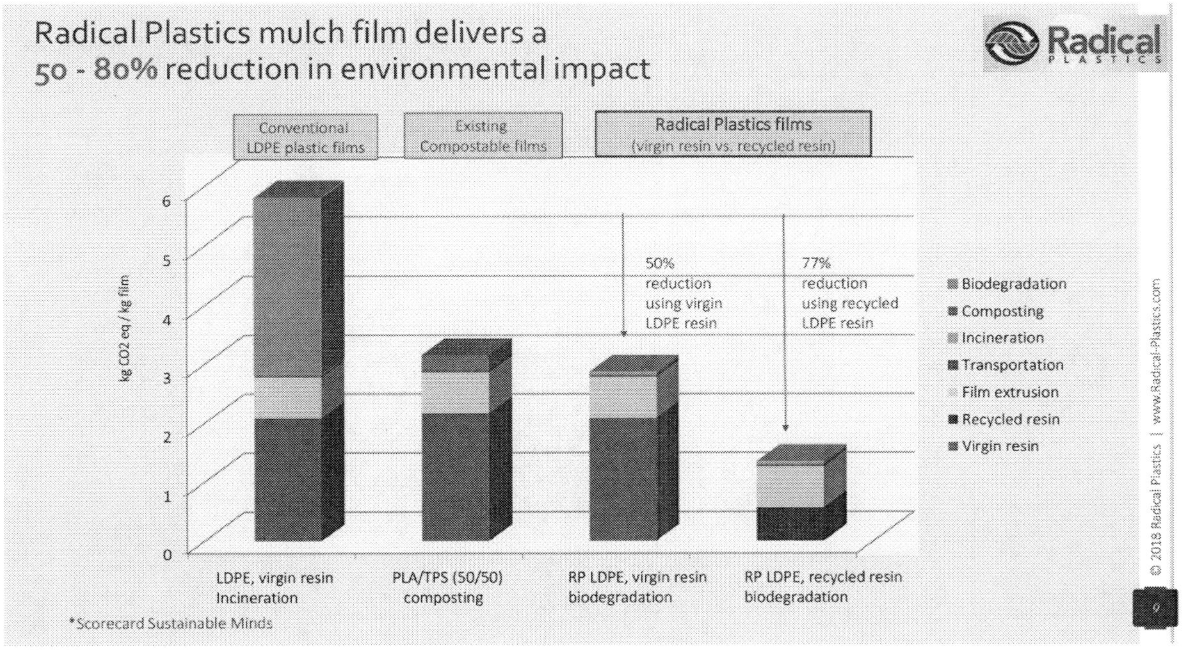
Radical Plastics mulch film delivers a
50 - 80% reduction in environmental impact
Radical PLASTICS
Conventional LDPE plastic films
Existing Compostable films
Radical Plastics films (virgin resin vs. recycled resin)
kg CO2 eq / kg film
0
1
2
3
4
5
6
50% reduction using virgin LDPE resin
77% reduction using recycled LDPE resin
Biodegradation
Composting
Incineration
Transportation
Film extrusion
Recycled resin
Virgin resin
LDPE, virgin resin Incineration
PLA/TPS (50/50) composting
RP LDPE, virgin resin biodegradation
RP LDPE, recycled resin biodegradation
*Scorecard Sustainable Minds
© 2018 Radical Plastics | www.Radical-Plastics.com
9

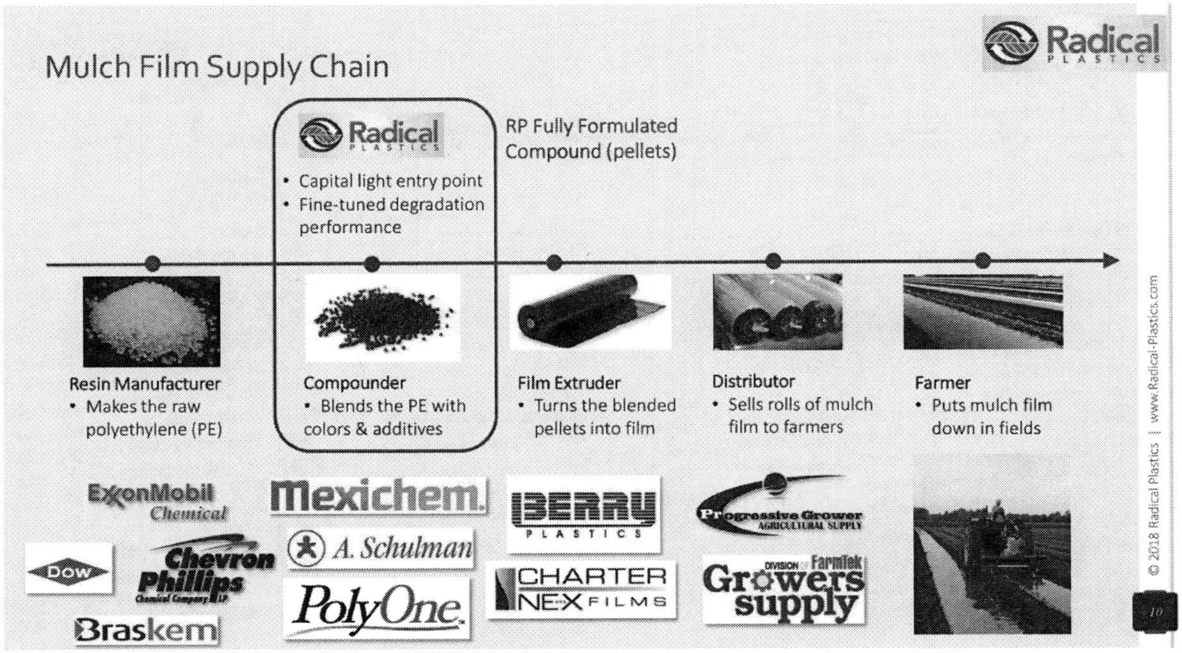
Mulch Film Supply Chain
Radical PLASTICS
RP Fully Formulated Compound (pellets)
• Capital light entry point
• Fine-tuned degradation performance
Resin Manufacturer
• Makes the raw polyethylene (PE)
Compounder
• Blends the PE with colors & additives
Film Extruder
• Turns the blended pellets into film
Distributor
• Sells rolls of mulch film to farmers
Farmer
• Puts mulch film down in fields
ExxonMobil Chemical
DOW
Chevron Phillips Chemical Company LP
Braskem
Mexichem
A. Schulman
PolyOne
BERRY PLASTICS
CHARTER NEX FILMS
Progressive Grower AGRICULTURAL SUPPLY
Growers supply
© 2018 Radical Plastics | www.Radical-Plastics.com
10

Radical Plastics allows for significant margin expansion

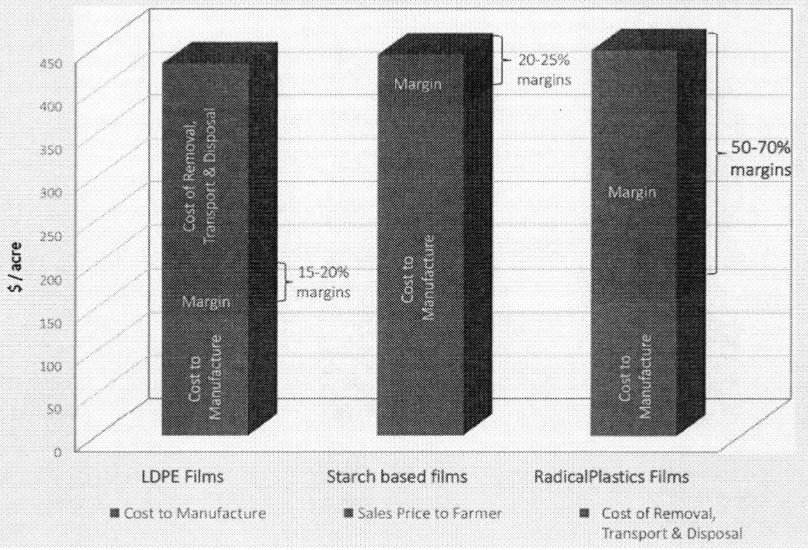

11

State of the Technology:

Compound / Film Produced To Date:

- Approximately 200 meters of film produced – U Mass Lowell and U.S. Army Natick Soldier RD&E Center
- Lab scale formulation of compound (16mm and 26mm twin screw extruders)
- Film extrusion (24" and 46" dies)

Preliminary Testing Completed:

- Mechanical properties
    - Tensile strength, elongation ASTM D882
    - Tear strength (ASTM D1922)
    - Dart Impact (ASTM 1709)
- Density (ASTM D1505)
- Optical properties (color, transmission, ASTM D6290)
- Chemical properties:
    - Film composition analysis after accelerated and non-accelerated conditions (ATR FTIR)
    - Fine mineral matter (ICP analysis)

Additional Testing In-progress:

- Biodegradation
    - Respirometry (USDA)
- EN17033:2018: Plastics - Biodegradable mulch films for use in agriculture and horticulture - Requirements and test methods

12

Target Physical Properties Have Been Achieved

Radical PLASTICS

Preliminary Testing Completed:

- Mechanical properties
  - Tensile strength, elongation ASTM D882
  - Tear strength (ASTM D1922)
  - Dart Impact (ASTM 1709)

Property	Same as LDPE
Tensile strength, GPa, max	0.3
E Modulus, GPa	0.8
Elongation at break, %	250-300
Density, g/cm3	0.92
Melting temperature, °C	122
Glass transition temperature, °C	-25

Readiness to Biodegrade Has Been Confirmed

Radical PLASTICS

- Films were installed outdoors for 2.5 months – summer in MA
- Films were removed and kept in the dark for ~ 1 year
- Films had degraded into a waxy brittle material with a Carbonyl index, as measured by ATR FTIR, with a ratio of 1710/720 1/cm) = ~0.9 which confirms that the material is approaching the requirements for biodegradability

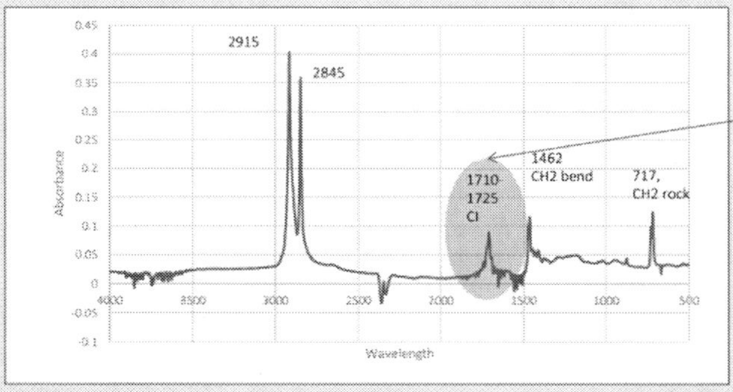

Composition change is demonstrated

Requirements for EN17033:2018 Certification

Radical PLASTICS

Table 1. Testing requirements and criteria for biodegradable plastic mulches, as outlined in EN 17033 (Sections 5 and 6).

Criterion	Test Method [1]
I. Constituents	
• Concentration limits for heavy metals [2]	EN 17294-2
• No hazardous substances of "very high concern" (<0.1%) [3]	
• Loss of ignition at 550°C (≥60%) [4]	
II. Biodegradation	
• ≥90% conversion of mulch's carbon into CO_2 within 2 years under ambient soil conditions [5]	ISO 17556 [6]
III. Ecotoxicity	
• No acute ecotoxicity to plants (≥90% of germination rate and plant growth achieved compared to mulch-free soil)	OECD 208 [7]
• No acute ecotoxicity to invertebrates (earthworms: *Eisenia fetida* or *E. andrei*; <10% difference in mortality rate and biomass amount compared to mulch-free soil [8])	ISO 11268-1,and -23 [7]
• No acute toxicity to microorganisms (nitrification inhibition of bacteria: nitrification should be ≥80% of that achieved for BDM-free soil)	ISO 15685 [7]
IV. Dimensional, mechanical, and optical properties [9]	
• Tensile stress at break ≥18 MPa (MD); ≥ 16 MPa (TD) [10]	ISO 727-1 and -3
• Light transmission of ≤3% (only for black and opaque films)	
V. Miscellaneous	
• Proper labeling; preparation of a test report by manufacturer	
• Surface area should not decrease <10% during deployment (if best practices are used).	Annexes G and H of EN 17033

Certification Mark from DIN CERTCO:

BIODEGRADABLE IN SOIL DIN Geprüft

Timing:

- 3-6 months from time of submission

Pricing:

- \$4500-\$5800 for initial sample
- \$1500-\$3000 for formulation modifications

15

Rapidly growing global market for plastic mulch

Radical PLASTICS

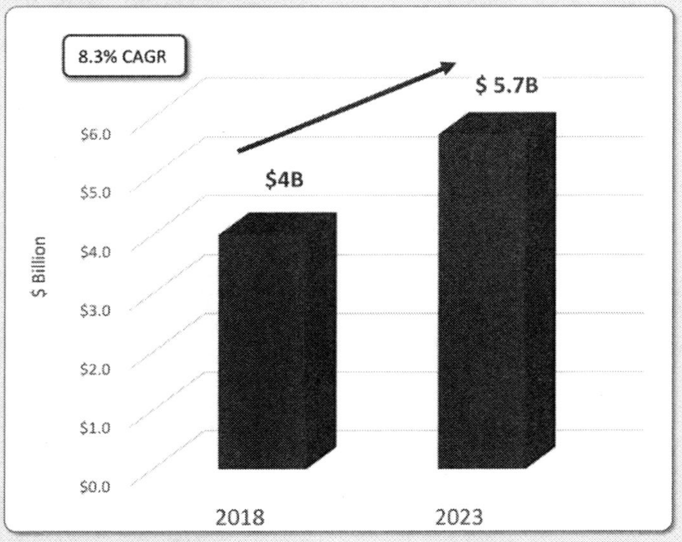

List of farmers waiting for Radical Plastics mulch film:

- Langwater Farm (organic), N. Easton, MA
- Marini Farm, Ipswich, MA
- Smolak Farm, N. Andover, MA
- Ladybug farm produce, Hubbardston, MA
- Denison Farm (organic), Schaghticake, NY
- Cold Spring Brook Farm, Berlin, CT
- Scott's Yankee Farmer, East Lyme, CT
- Easy Pickin's, Enfiled, CT
- Cicarelli Farm, Northford, CT,
- Stone Garden Farm, Shelton, CT

16

Radical Plastics Roadmap to Success

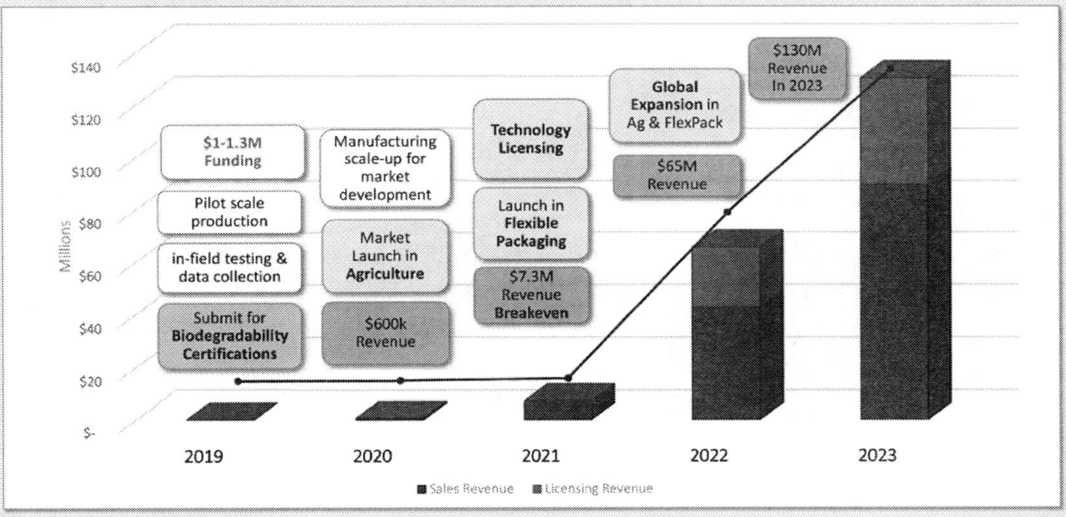

17

Financial Plan* - $1.3M investment

5-Year Financial Projection

	2019	2020	2021	2022	2023
Sales (Revenue)	$0	$600,000	$6,750,000	$42,750,000	$90,000,000
Sales (Revenue - Agriculture)		$600,000	$6,750,000	$15,000,000	$30,000,000
Sales (Revenue - Packaging)		$0	$0	$27,750,000	$60,000,000
Licensing revenue			$572,000	$22,880,000	$40,612,000
Total Revenue	$0	$600,000	$7,322,000	$65,630,000	$130,612,000
COGS (Cost of Goods Sold)	$0	$450,000	$4,275,000	$24,225,000	$51,000,000
Gross Profit	$0	$150,000	$3,047,000	$41,405,000	$79,612,000
Margin		25%	42%	63%	61%
Operating Expenses					
SG&A	267,400	469,800	571,000	2,625,200	5,224,480
R&D	$153,000	$55,000	$270,000	$1,710,000	$3,600,000
Total Operating Expenses	$420,400	$524,800	$841,000	$4,335,200	$8,824,480
EBITDA	-$420,400	-$374,800	$2,206,000	$37,069,800	$70,787,520
DA	$23,000	$35,200	$46,400	$57,600	$68,800
Less taxation	$0	$0	$391,860	$11,103,660	$21,215,616
Net profit for the year	-$443,400	-$410,000	$1,767,740	$25,908,540	$49,503,104
Less CapEx spent	$115,000	$61,000	$56,000	$56,000	$56,000
Less Working Capital Increased	$0	$175,000	$1,757,833	$13,043,000	$15,292,833
Net Cash Flow Before Financing	-$558,400	-$646,000	-$46,093	$12,809,540	$34,154,271
Starting Cash	$0	$741,600	$95,600	$49,507	$12,859,047
Plus Financing (Funding)	$1,300,000				
Net Cashflow - Ending Cash	$741,600	$95,600	$49,507	$12,859,047	$47,013,317

Use of Funds

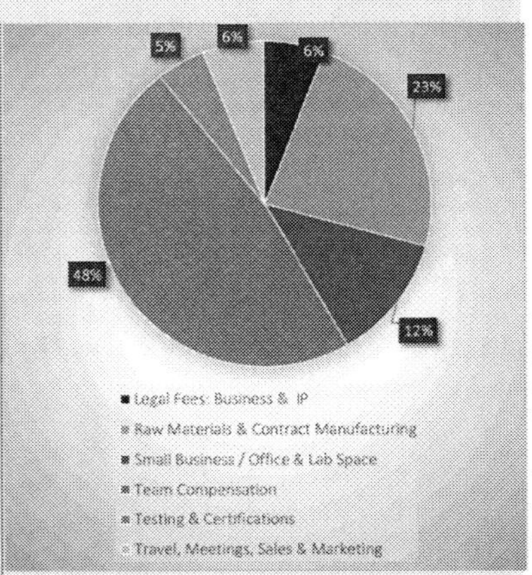

*Financial projections do not include any new grants, business plan competition awards or JDA funding, which are all being aggressively pursued.

18

Flexible Packaging is a $150B Industry
with a very urgent and very visible problem

19

Radical PLASTICS

WE CAN DO BETTER!

REUSE
REDUCE
RECYCLE
REDESIGN

...WE KNOW HOW

20

Team Members & Relevant Experience

Decades of success in plastics & bioplastics industry

Yelena Kann, PhD

- PhD, Chemistry & MS, Polymer Engineering & Technology
- 25+ years in industry and academia, Over 20 patents and patent applications in Bioplastics, Polymer Chemistry & Color Technology, Over 30 published papers
- Work experience: **Metabolix Bioplastics**, Sabic (formerly GE Plastics), EInk
- Extracurricular: Skiing, Hiking, Rock climbing, Agility dog training

Kristin Taylor

- MS, Plastics Engineering
- 20+ years in Market Development for commodity, engineering and biobased / biodegradable plastics
- Work Experience: ExxonMobil, **Metabolix Bioplastics**, Sabic (formerly GE Plastics), Huntsman Corporation
- Extracurricular: Long distance running, 17 marathons including the 6 World Majors (2018)

Advisors: Extensive business & entrepreneurial experience

Rick Eno

- Former CEO, **Metabolix Bioplastics**, Former CEO and Board Member, BioAmber
- 25+ years of management experience, Senior Partner of a leading global management consultancy. Industry expertise in energy, chemicals, materials and biotechnology.
- B.S. in Chemical Engineering, MBA and Chartered Financial Analyst (CFA)

Sam Nejame

- Founder web based B2B and B2C companies. Project management and engineering at Parsons Corp and in Pharma/Biotech, Growth oriented management consultant with 20+ years experience.
- B.S. in Chemical Engineering, WPI, MBA, Northwestern Univ. Kellogg School of Management

21

Collaborators: Expertise in multiple areas (soil science, horticulture, polymers)

Allison Flynn, Bioproducts Research Division, USDA

Dr. Carol Miles, Professor, Department of Horticulture, Washington State University

Dr. Kate Lajtha, Professor, Biogeochemistry, Department of Crop & Soil Sciences, Oregon State University

Dr. JoAnn Ratto, Polymer Research Engineer, U.S. Army Natick. Soldier Research, Development and Engineering Center (NSRDEC) Natick, MA

Dr. Marion Brodhagen, Associate Professor, Department of Biology, Western Washington University

Anusuya Rangarajan, Director, Cornell Small Farm Program

22

Radical PLASTICS
ALLIANCE TO END PLASTIC WASTE
Dow Chemical CEO: 'I've never seen the industry more aligned' around tackling plastic waste
DSM
NOVA Chemicals
DOW
P&G
PolyOne
Berry
TOTAL
VEOLIA
SUEZ
lyondellbasell
ExxonMobil
OxyChem
CLARIANT
SCG
Sasol
"We strongly support the aim to reduce plastic waste in the environment. We are co-founding the Alliance to End Plastic Waste, because we want to drive and promote solutions that will effectively help solve the world's plastic waste problem."
- Martin Brudermüller, CEO of BASF SE
www.endplasticwaste.org
© 2018 Radical Plastics | www.Radical-Plastics.com
23

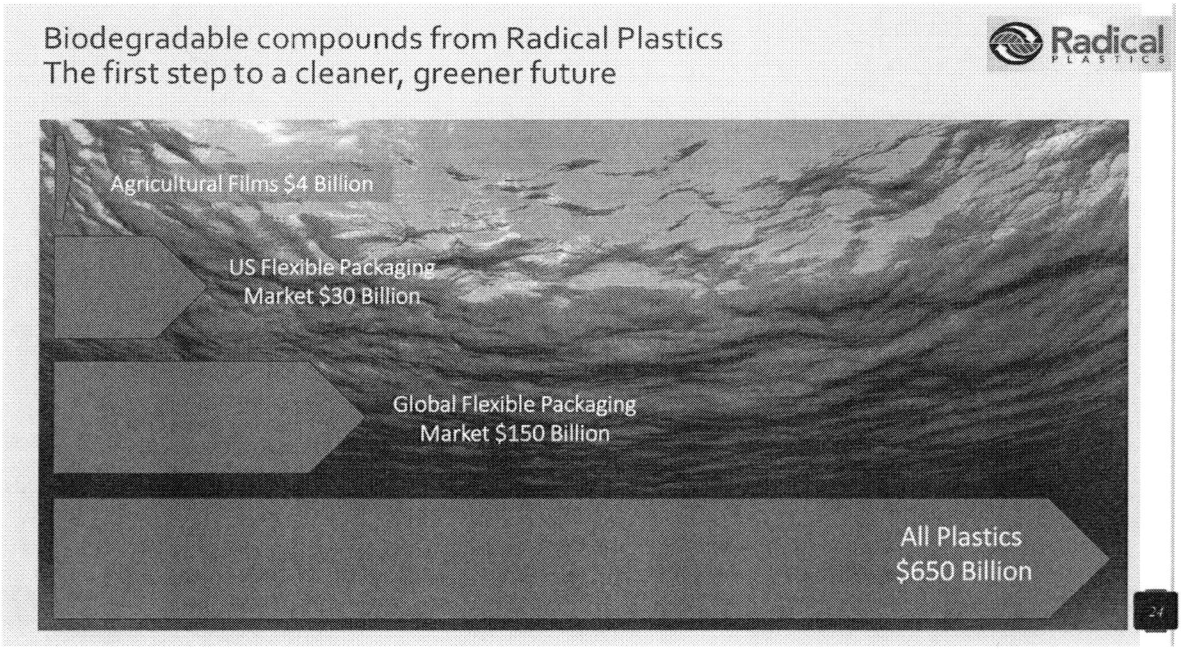
Biodegradable compounds from Radical Plastics
The first step to a cleaner, greener future
Radical PLASTICS
Agricultural Films $4 Billion
US Flexible Packaging Market $30 Billion
Global Flexible Packaging Market $150 Billion
All Plastics $650 Billion
24

Radical PLASTICS

a **radical** solution to **plastic** pollution

Thank you

Kristin Taylor, Co-founder / CEO
ktaylor@radical-plastics.com
(617) 909-9032

Dr. Yelena Kann, Founder / CTO
ykann@radical-plastics.com
(617) 320-9663

www.Radical-Plastics.com

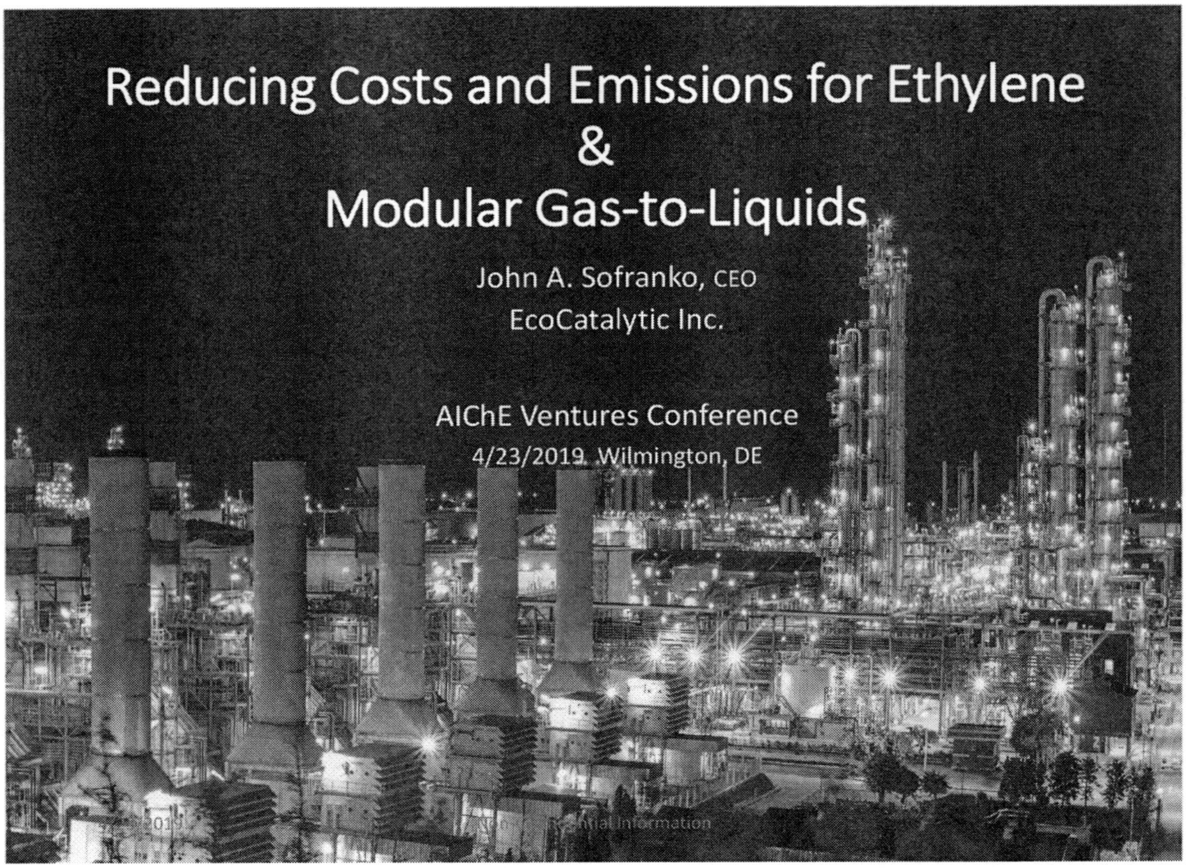

EcoCatalytic: Business, Products, Services

EcoCatalytic Technologies | **EcoCatalytic Inc.**

Gov't & Private Funding → Research and Development → IP Portfolio → Scale-Up → Licensing

World Scale Ethylene Plants

$11.1 MM Funding – TRL 6

Modular Gas-to-Liquids

$1.2 MM Funding – TRL 3

Management Team

Two Companies, One Mission: *develop & commercialize hydrocarbon conversion technologies for more sustainable chemical & fuel processes*

EcoCatalytic Technologies

Founded: 2009 (as Bio2Electric, LLC; rebranded in 2014)

IP Licenses →

← Tech-to-Market Activities

EcoCatalytic Inc.

Founded: 2018

Dr. Elena Y. Chung, CEO
Thesis in Chemical Looping at OSU
Numerous patents and publications

Dr. Carol S. Lee, Managing Partner
Research & management at Mobil Chemical

Dr. John A. Sofranko, COO
Pioneering discoveries in hydrocarbon oxidation since 1980s; >60 area patents
Executive management ARCO, Lyondell, AIChE

Dr. John A. Sofranko, CEO

Dr. Elena Y. Chung, Dir. Research & Engineering

Dr. Larry Evans, Business Advisor
>50 years of entrepreneurial success (Aspen Tech)

James Tarangelo, Business Advisor
>50 years of executive management, M&A (ARCO, Sorgenti Investment)

Challenges	Solutions
World Scale (1-2 MM t/yr.) Ethylene Production	
Steam Cracking of Ethane	**Ethane Oxidative Dehydrogenation**
High energy consumption for endothermic conversion; high CO_2 emissions from fuel combustion	Chemical looping oxidative dehydrogenation is net <u>exporter</u> of energy; low CO_2 emissions
NO_x emission control; constraints in EPA non-attainment areas	Flameless combustion produces zero NOx
High capital and operating costs due to multiple reactor furnace complex	One fluid bed reactor for up to 1.5 MM tpy ethylene, similar to FCC reactors
Modular Gas-to-Liquids	
Modular Fischer-Tropsch	**Modular OCM and ODH**
High capital to unit production volume ratio	Electrogenerative reactors leverage modular SOFC manufacturing systems Novel modular chemical looping system under development
On-site shale gas production sites lack infrastructure	Minimal site requirements (no air separation)
Multiple products require further refining	One fungible product for on-site use

Technologies and IP

- Methane oxidative coupling (OCM) to ethylene and water
- Ethane oxidative dehydrogenation to (ODH) ethylene and water [**Core technology**]
- Naphtha to ethylene (olefins)
- Modular Gas-to-Liquid reactors for natural gas to fuels and electricity/power

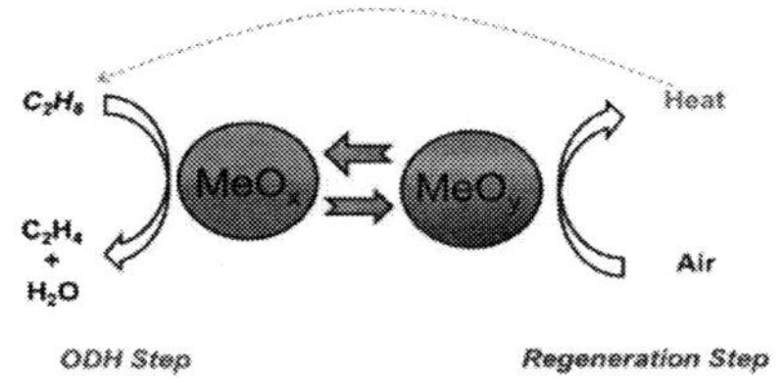

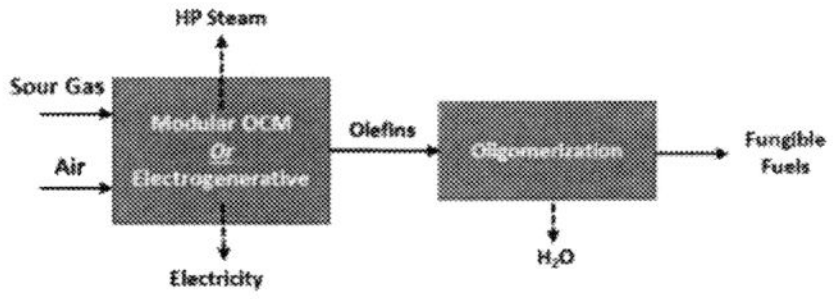

TITLE	APPLICATION	STATUS
REACTOR, PROCESS, AND SYSTEM FOR THE OXIDATION OF GASEOUS STREAMS	US2017-0247803	Pending, Published
OXYGEN TRANSFER AGENTS FOR THE OXIDATIVE DEHYDROGENATION OF HYDROCARBONS AND SYSTEMS	US 10,138,182	Allowed
SYSTEMS FOR PROMOTING ENDOTHERMIC CONVERSIONS WITH OXYGEN TRANSFER AGENTS	WO 2016-209811 A1	Claims Allowed Not Published
ENHANCED OXYGEN TRANSFER AGENT SYSTEMS FOR OXIDATIVE DEHYDROGENATION OF HYDROCARBONS AND METHODS OF USING THE SAME	WO 2018/005456 A1	Pending, Published
OXIDATIVE CO-CRACKING OF HYDROCARBONS	WO 2018/075713 A1	Pending, Published
METHODS AND SYSTEMS FOR OXIDATIVE DEHYDROGENATION OF HYDROCARBONS TO OLEFINS	WO 2018/208907 A1	Pending, Published
Various catalyst and process innovations	4 Provisional Patents	Unpublished

Markets & Competitive Landscape

Small Scale Modular Gas-to-Liquids

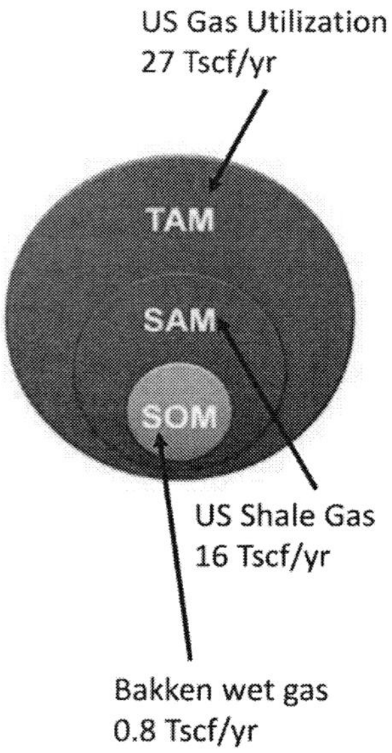

Company	Product (Technology)	Scale: Small <1MMscfd	Scale: Medium 1-10 MMscfd	Scale: Large >10MMscfd	Status
CompactGTL	Diesel+ (FT)			✓	Commercial ↑
Greyrock	Diesel+ (Direct FT)		✓	✓	
SGC Energia	Diesel+ (FT)		✓	✓	
Velocys	Diesel+ (FT)	✓	✓	✓	
Aum Energy	DME		✓		
EFT	Diesel+ (FT)		✓	✓	
GasTechno	MeOH+ (POX)	✓	✓		
Infra Technology	Diesel+ (Direct FT)	✓	✓		
Maverick Synfuels	MeOH+	✓	✓		
Oberon Fuels	DME, MeOH	✓	✓		
Primus Green Energy	MeOH, Gasoline		✓	✓	
Proton Ventures	Ammonia	✓			
R3Sciences	MeOH	✓			
TIGAS	Gasoline (FT)		✓	✓	
Biofuels Power	Diesel+ (FT)		✓		
Synfuels	Ethylene, gasoline (OP)			✓	
Gas2	Diesel+ (FT)	?	?	?	
Greenway IE	Diesel+ (FT)		✓	✓	
Methion	MeOH+ (MSA)	✓	✓	✓	
Verdis	Diesel+ (Direct FT)	✓	✓		
Siluria	***Ethylene, gasoline (OCM)***	*?*	*?*	✓	
EcoCatalytic	***Ethylene, gasoline (OCM)***		✓	✓	
					R&D

Financing: Investment Sought and Use of Funds
B-Round 2020-2021

Technology	e-GTL	Redox OCM/ODH
Purpose/Use of Funds	Scale up SOFC reactor system to 10kW stack – Commercial fuel cell partner identified	Scale up novel modular OCM/ODH reactor system; 10 kg/D Engineering partners identified
Anticipated Gov't	$1.1 MM	$1.1 MM
Founder's Funding	$0.3 MM	$0.3 MM
Additional Private Funds	$ 2.2 MM	$3 MM

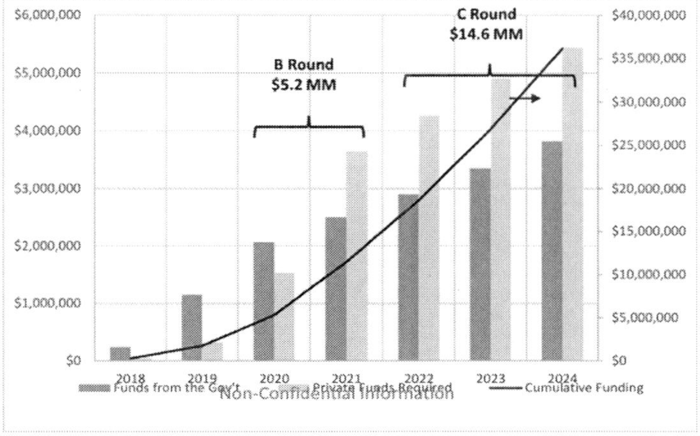

Summary and Contact Information

World scale ethylene via chemical looping

- Pilot unit funded via DOE/EERE contract ($2 MM Gov't/ $2 MM company(s) cost share)
- Growing patent platform
- Licensing and investment discussions in advanced stages

Modular wet gas Gas-to-Liquids

- Two parallel process approaches
- Phase I SBIR/STTR proof of concept completed
- Applying for additional phase II funding
- Technology partnership discussions on-going for scale-up
- ***Seeking private investment funds up to $5.2 MM for B round***

Contact:
John A. Sofranko, CEO
EcoCatalytic Inc.
jasofranko@ecocatalytic.com
609 865 2309
561 Virginia Rd.
Concord, MA 02493

End of Presentation

Copprint

The basic building block of printed electronics

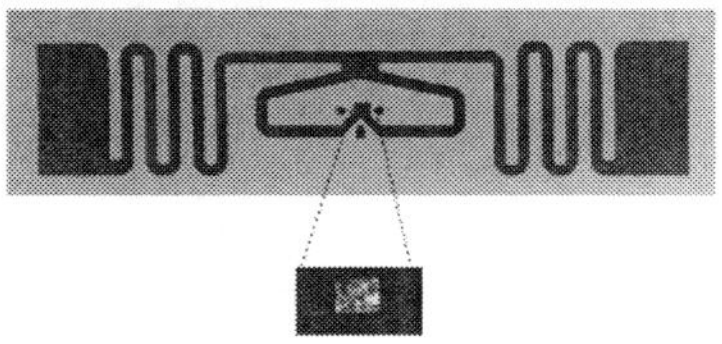

Dr. Ofer Shochet
CEO & Founder

ofer@copprint.com

A breakthrough in conductive inks

Nano Copper technology – Less expensive, More sustainable, Faster to produce

Enabling the electronics industry to adopt printed electronics

- Current alternatives: Printing silver inks – expensive and eco damaging ; Etching – eco damaging
- Our award-winning technology will disrupt the industry (LOPE-C Startup award - 2016; Printed Electronics AWARDS 2017)
- Strong IP and manufacturing trade secrets
- Enthusiastic engagement from all customers & strategic partners (Mühlbauer High Tech International; Henkel; hp)
- High volume immediate market opportunities in RFID and NFC Antennas, PV and enormous future market

Copprint was founded in 2016 – working two years in a garage mode

CTO: Michael Grouchko, PhD (Chem)

- 2012-2015 - Clear-Jet: Co-founder and CTO
- 2003-2013 – Hebrew University
- World expert in conductive nano-metal inks
- Commercialized silver nano-inks for XJet Solar and NanoDimension

CEO: Ofer Shochet, PhD (Phys)

- 20+ years in high-tech
- Past 10 years – Executive @ 3D printing (Stratasys – EVP Products, responsible for all 3D printing inkjet development, Velo3D - BOD)
- Founder of multiple startups and an executive in public companies (Navajo was acquired by salesforce.com @2011)
- Israel's National security award

XJET, האוניברסיטה העברית בירושלים THE HEBREW UNIVERSITY OF JERUSALEM, NANO DIMENSION, ClearJet

Stratasys, OBJET, SiliconGraphics Computer Systems, Vigil, VELO3D, VERINT, KREDO, Navajo Systems, RoboUniverse

11 employees
4PhDs, 3 MSc, 2 Eng

Investors: Henkel, hp

Creating a conductive pattern:

Subtractive methodologies

- Etching of copper/aluminum foils(~$70B market)
- Die cutting of aluminum foils

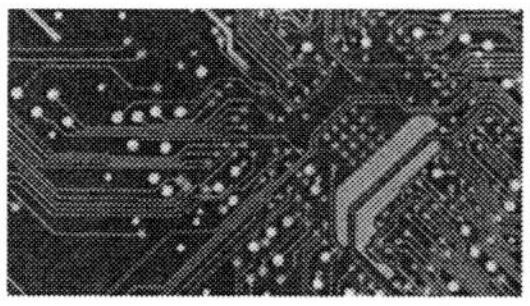

Additive methodologies

- Silver based inks (~$2B market)
- Copprint's Nano-Copper ink New

SUBSTRATE FREEDOM
LOW COST
SIMPLICITY/COMPLEXITY OF PROCESS
ENVIRONMENTAL IMPACT
ROBUSTNESS
CONDUCTIVITY

Conductive Copper inks were the holy grail for the research community

For many years people have attempted to develop copper inks

However, oxidation prevented significant conductivity

- Much effort devoted to develop a process to overcome oxidation – failed in large scale

The Copprint Approach - Patented chemical sintering agent is added to Copper Nano-Particle

- During drying and sintering a chemical reaction is induced and prevents oxidation and enable fast sintering

Finally you can easily print Conductive Copper Ink

Screen printing (link)

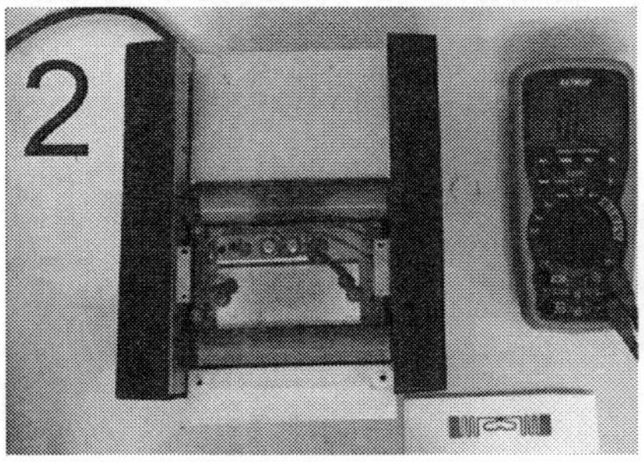

Sintering using a $200 office laminator (link)

Conductivity = 30% of bulk Copper
Sheet resistance = 2.3×10^{-3} Ω/□/mil
Resistivity = 5.6×10^{-8} Ω×m

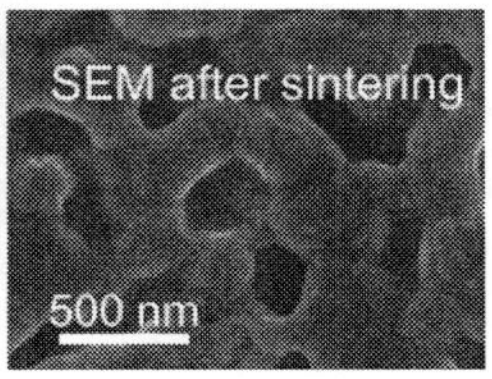

- Patented Ink, process and article (granted patent + PCT process)
- Trade secrets for efficient nano Copper manufacturing – substantially cheaper than nanoCU vendors
- Can be implemented for screen printing, flexographic, gravure, inkjet

Products

- Our products portfolio is focused on Screen-printing.
  - LF-300 – paste for paper substrate (TRL level 8)
  - LF-350 – paste for PET substrate (TRL level 6-7)
  - LF-390 – paste for PI substrate (TRL level 5-6)
  - LF-380 – paste for HJT PV cells (TRL level 4-5)

- Products vision
  - Most silver inks can be replaced with nano-Copper based inks
  - Minor challenges supporting additional printing/dispensing techniques such as Inkjet, Flexo, Gravure. We already performed successful POC with inkjet and dispensing.

Comparison to silver inks

- Conductivity as top silver inks
- 3-5 x Cheaper than silver inks
- Short sintering/curing process
- Almost any substrate
- Environmental friendly
- Short, simple and cheap process

Lower sheet resistance = better conductivity

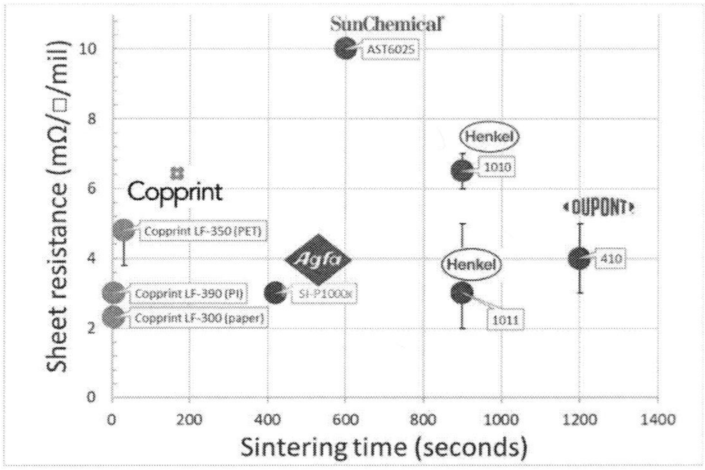

Comparison matrix

	Copprint	Silver inks	Aluminum Die cutting	Etching	Photonic sintering of copper oxide	Carbon inks
Cost of material	✓	High cost – Raw silver >$550/kg				
Paper substrate	✓			Only acid resistant substrates		
Cost of Process	✓				~$0.5M – cost per curing system	
Accuracy	✓		Very inaccurate due to process			
Environmental friendly process	✓			Problematic due to acids		
Compostable	✓	Silver is toxic to the environment	Must be recycled	Must be recycled		
Sintering time	✓	Typical sintering of minutes to an hour				
Robustness	✓				Not stable for a long period ; Complex process	
High conductivity	✓					1/1000 in conductivity
Homogeneity	✓				Only thin layer next to the surface is sintered well	
Digital production	✓ Inkjet as well		Limited patterns	Long turnaround		

Markets

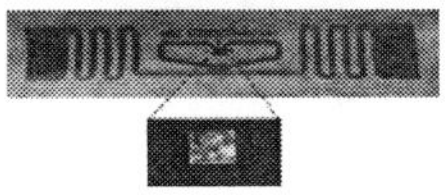

Paste for printing RFID antennas on paper

- Estimated 2025 market: Trillion units & $2B TAM
- Collaborating with world leader RFID equipment manufacturer Mühlbauer High Tech International
- **Cheaper** and **more sustainable** relative to silver ink and to etched antennas

Paste for Heterojunction Photovoltaic (PV) cell metallization

Demonstrated similar efficiencies as top silver pastes with 3x cheaper cost.

Reduced costs of HJT PVs to be similar to PERC cells - revolutionizing the PV market

Evolving new applications: Printing PCBs and flexible electronics (TAM > $1B)

Copprint

Antenna Printing Solution (APS)

- Muhlbauer is the world leader in RFID equipment (>$0.5B annual rev)
- >90% of RFIDs are made by Muhlbauer machines
- Exclusive strategic partnership

Mühlbauer
High Tech International

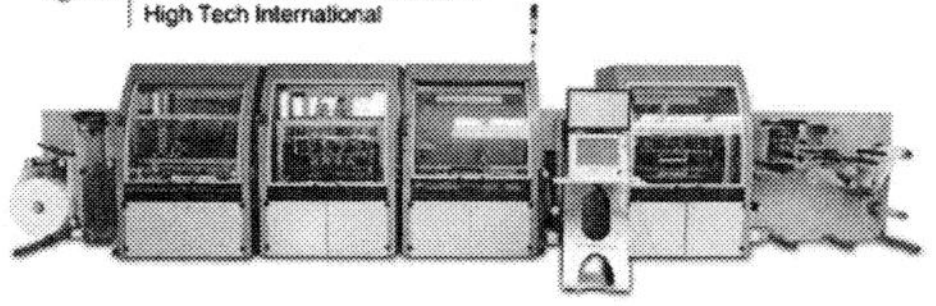

- Manual printing was already approved by top customers
- APS is ready

Copprint

Domination

- First to market – Achieve exclusivity and relationship with OEMs
  - E.g. Muhlbauer for RFID , Fortune 500 OEM for the PV market
- Strong IP filed for ink formulation & process + more to come
  - Granted patent in one territory and PCT worldwide
- Trade secrets for low cost, efficient nano Copper ink manufacturing
  - Substantially cheaper than state of the art
  - Aiming for high GM

Copprint

Financing

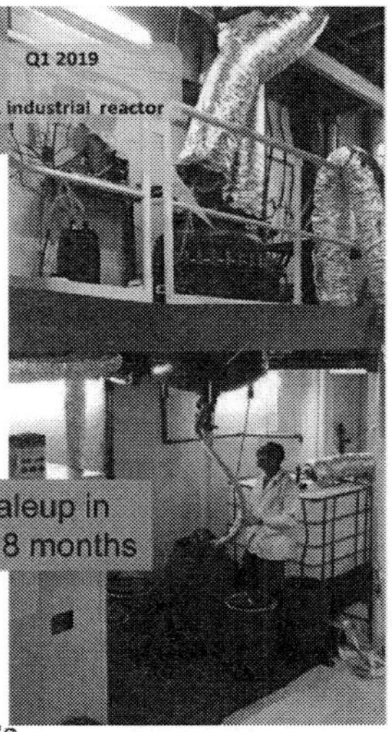

Key achievements since series A round (Nov 2017):

- Reduced all scaleup technological risks – only engineering risks remain
- Approved cost structure and projected profitability.
- 2 founders → 11 employees (4 PhDs, 3 MSc, 2 Eng).
- Applications: – validated RFID, PV and flexible electronics products Mühlbauer
- Business: Signed strategic agreement with commitment on ink tons

Next steps:

- New production facility
- Management & Infrastructure scaleup
- Large scale delivery of LF-300 conductive ink for RFID antennas
- Additional applications & IP focus: PV, Sensors printing, Heaters printing, Flexible PCBs
- Additional printing techniques: Dispensing, Flexo and inkjet printing
- Support Initial revenues in 2019 and solid growth forward

Summary - A breakthrough in conductive inks

Nano Copper technology – Less expensive, More sustainable, Faster to produce

Enabling the electronics industry to adopt printed electronics

- Current alternatives: Printing silver inks – expensive and eco damaging ; Etching – eco damaging
- Our award-winning technology will disrupt the industry LOPE-C Startup award - 2018; Printed Electronics AWARDS 2017
- Strong IP and manufacturing trade secrets
- Enthusiastic engagement from all customers & strategic partners Mühlbauer Henkel hp
- High volume immediate market opportunities in RFID and NFC Antennas, PV and enormous future market

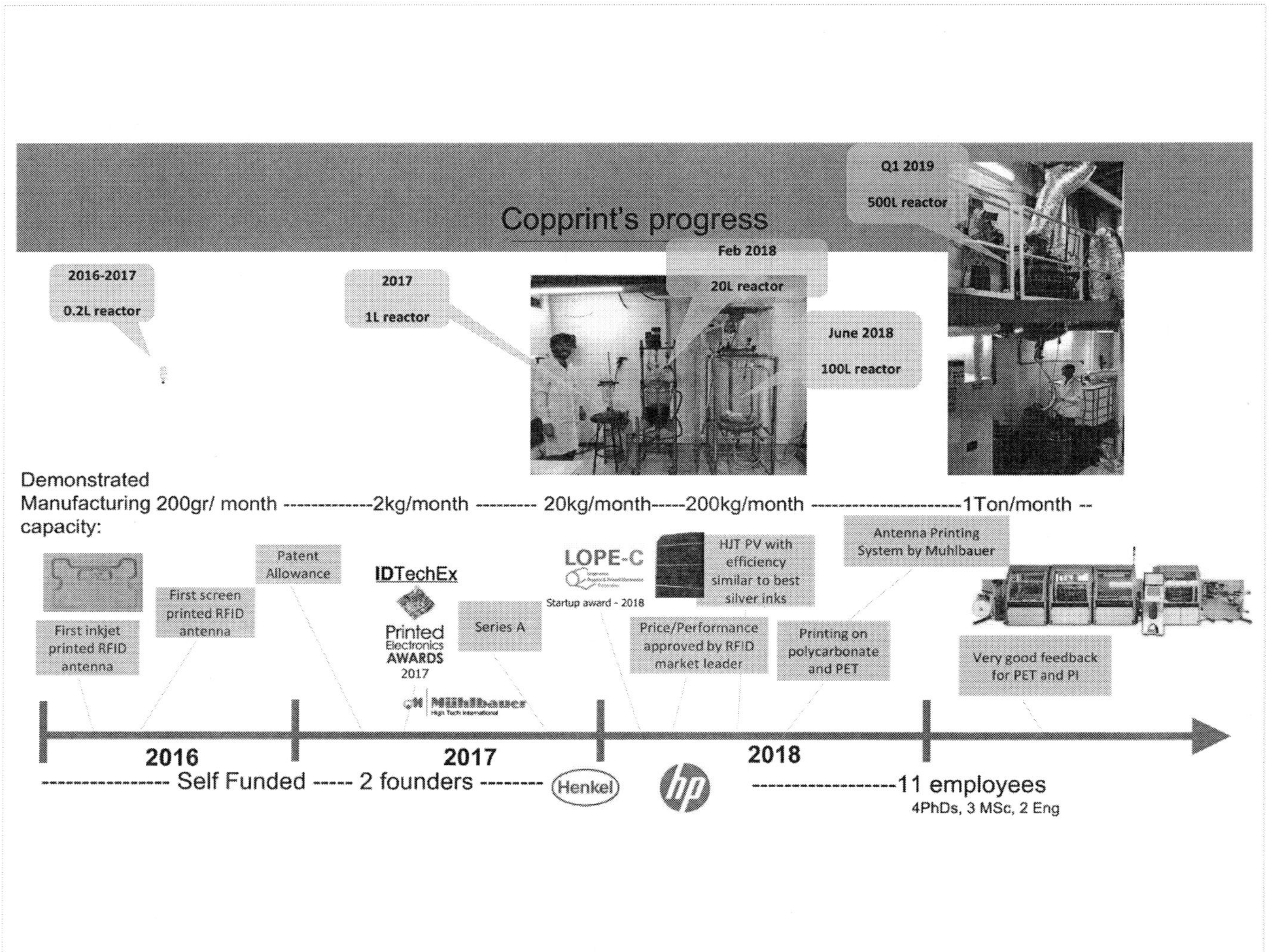
Copprint's progress
2016-2017
0.2L reactor
2017
1L reactor
Feb 2018
20L reactor
June 2018
100L reactor
Q1 2019
500L reactor
Demonstrated Manufacturing capacity: 200gr/ month -------------2kg/month --------- 20kg/month-----200kg/month ----------------------1Ton/month --
First inkjet printed RFID antenna
First screen printed RFID antenna
Patent Allowance
IDTechEx
Printed Electronics AWARDS 2017
Series A
LOPE-C
Startup award - 2018
HJT PV with efficiency similar to best silver inks
Price/Performance approved by RFID market leader
Printing on polycarbonate and PET
Antenna Printing System by Muhlbauer
Very good feedback for PET and PI
2016
2017
2018
---------------- Self Funded ----- 2 founders --------
Henkel
-------------------11 employees
4PhDs, 3 MSc, 2 Eng

Compact Membrane Systems

Compact
Membrane
Systems

Compact
Membrane
Systems

Industrial separations are a major user of energy

"Purification of propene and ethene alone accounts for 0.3% of global energy use, roughly equivalent to Singapore's annual energy consumption."

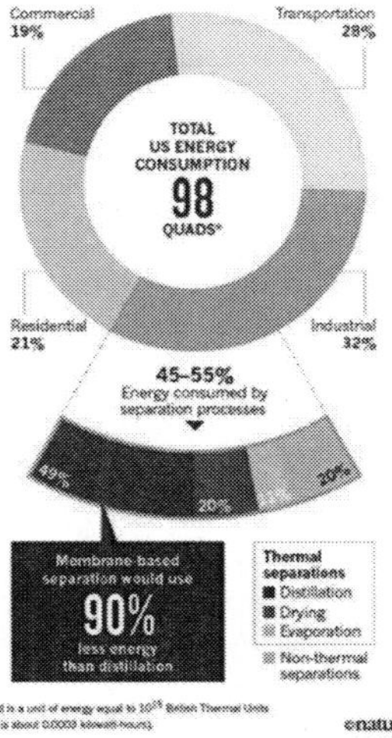

Compact
Membrane
Systems

Sholl, D.S., Lively, R.P., Nature, **532**, 435-437, 28 April 2016

2

New technology for olefin-paraffin separation

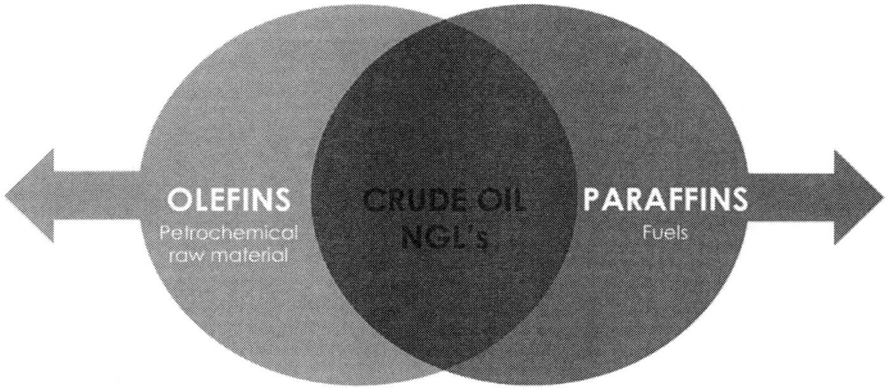

Compact Membrane Systems

3

Unlocking value with economically attractive separation at small scale

High capex, energy intensive	Low cost
Fixed, inflexible	Modular, flexible

Compact Membrane Systems

4

Optiperm™ membrane is a disruptive technology for separation

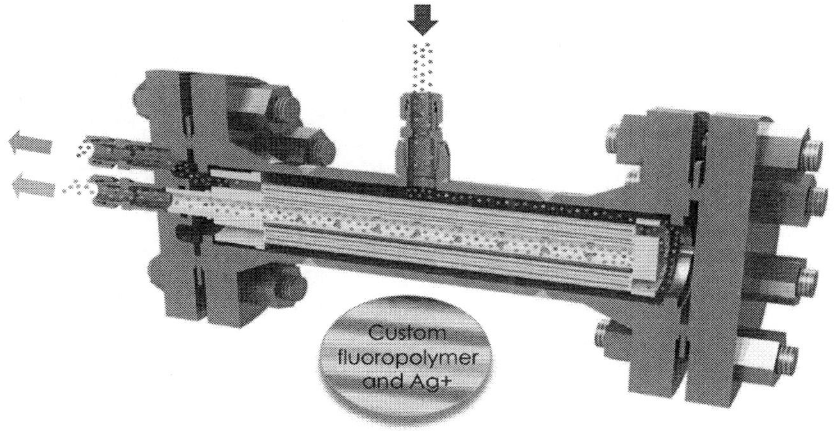

Compact Membrane Systems

5

>$5BN market opportunity

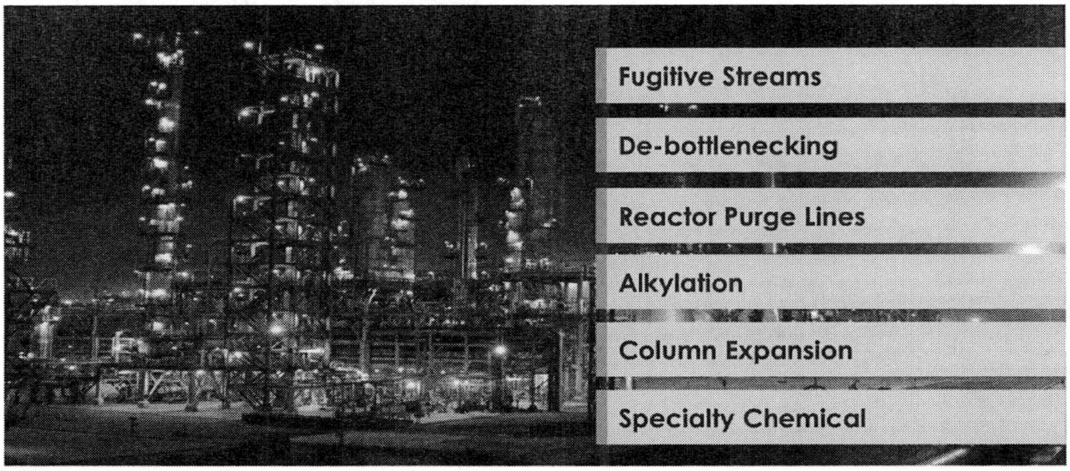

Compact Membrane Systems

6

Strong customer value is driving interest and trial

	IRR (%)	Pay back (Mos.)	Capex ($MM)	Opex ($MM)
Use Case 1: C_3 splitter expansion	**70%**	**16**	**$3.8**	**$1.8**
Use Case 2: Polypropylene reactor purge	**155%**	**8**	**$0.7**	**$0.2**

Compact Membrane Systems

7

On site test is significant proof point for industry

Delaware City Refining Company LLC

Results

- Successful, stable performance >50 days
- Financial models are consistent with observed field performance
- Field data matches laboratory data

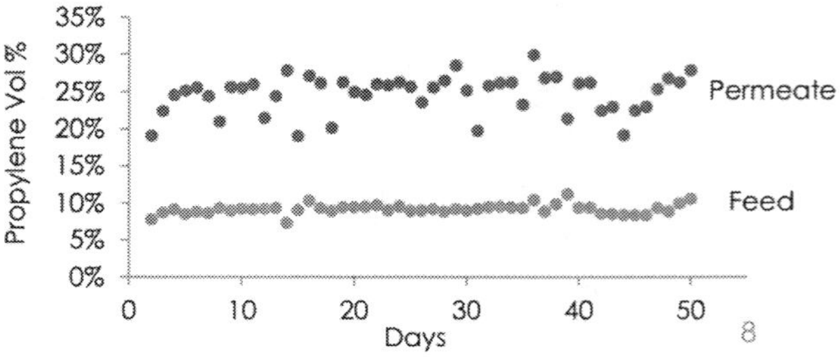

Compact Membrane Systems

8

Current pipeline

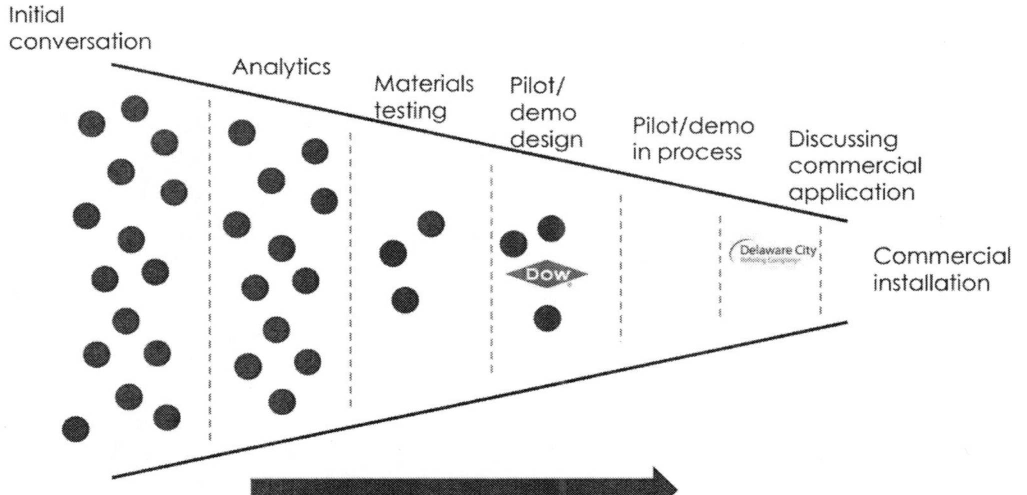

Compact Membrane Systems

9

Where are we currently?

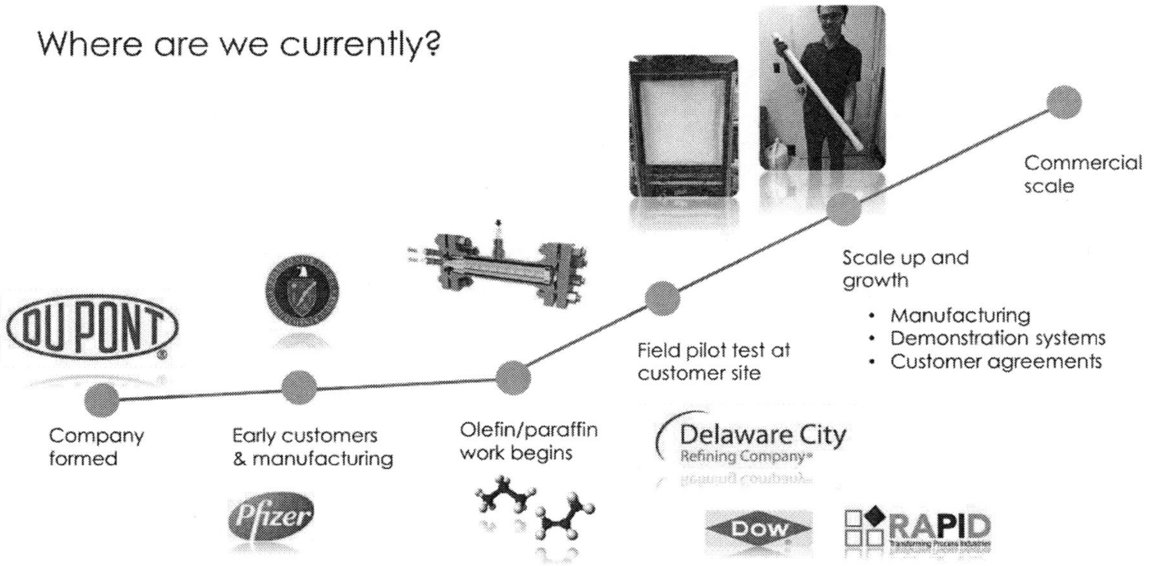

Compact Membrane Systems

10

Experienced team: Business, tech, industry

Erica Nemser, PhD
CEO

McKinsey&Company

Hannah Murnen, PhD
CTO

McKinsey&Company

William Charlton
Olefins

SUNOCO

Sudip Majumdar, PhD
Science

Stu Nemser, PhD
Founder

DUPONT

Ken Loprete
Product

Compact Membrane Systems

Compact Membrane Systems

11

We are looking for:

- Customers, pilots and demos
- Partnerships for acceleration
- Capital for scale-up

Compact Membrane Systems

12

Thank You

Hannah Murnen
Chief Technical Officer
hmurnen@compactmembrane.com

Erica Nemser
Chief Executive Officer
enemser@compactmembrane.com

AUTHOR INDEX